Christof Audretsch

Quorum sensing and biofilm formation in Staphylococcus aureus

Christof Audretsch

Quorum sensing and biofilm formation in Staphylococcus aureus

Analysing and simulating the importance of agr and sae for quorum sensing and biofilm formation in Staphylococcus aureus

Südwestdeutscher Verlag für Hochschulschriften

Impressum / Imprint
Bibliografische Information der Deutschen Nationalbibliothek: Die Deutsche Nationalbibliothek verzeichnet diese Publikation in der Deutschen Nationalbibliografie; detaillierte bibliografische Daten sind im Internet über http://dnb.d-nb.de abrufbar.
Alle in diesem Buch genannten Marken und Produktnamen unterliegen warenzeichen-, marken- oder patentrechtlichem Schutz bzw. sind Warenzeichen oder eingetragene Warenzeichen der jeweiligen Inhaber. Die Wiedergabe von Marken, Produktnamen, Gebrauchsnamen, Handelsnamen, Warenbezeichnungen u.s.w. in diesem Werk berechtigt auch ohne besondere Kennzeichnung nicht zu der Annahme, dass solche Namen im Sinne der Warenzeichen- und Markenschutzgesetzgebung als frei zu betrachten wären und daher von jedermann benutzt werden dürften.

Bibliographic information published by the Deutsche Nationalbibliothek: The Deutsche Nationalbibliothek lists this publication in the Deutsche Nationalbibliografie; detailed bibliographic data are available in the Internet at http://dnb.d-nb.de.
Any brand names and product names mentioned in this book are subject to trademark, brand or patent protection and are trademarks or registered trademarks of their respective holders. The use of brand names, product names, common names, trade names, product descriptions etc. even without a particular marking in this works is in no way to be construed to mean that such names may be regarded as unrestricted in respect of trademark and brand protection legislation and could thus be used by anyone.

Coverbild / Cover image: www.ingimage.com

Verlag / Publisher:
Südwestdeutscher Verlag für Hochschulschriften
ist ein Imprint der / is a trademark of
OmniScriptum GmbH & Co. KG
Heinrich-Böcking-Str. 6-8, 66121 Saarbrücken, Deutschland / Germany
Email: info@svh-verlag.de

Herstellung: siehe letzte Seite /
Printed at: see last page
ISBN: 978-3-8381-3841-1

Zugl. / Approved by: Würzburg, Universität Würzburg, Dissertation, 2014

Copyright © 2014 OmniScriptum GmbH & Co. KG
Alle Rechte vorbehalten. / All rights reserved. Saarbrücken 2014

Acknowledgements

First of all I would like to thank Prof. Dr. Thomas Dandekar and his work group at the Chair for bioinformatics at the biocenter in Würzburg for giving me the possibility to write this thesis although I was not the usual PhD student, engaged in studying Medicine in Tübingen at the same time. Furthermore I would like to thank him for supporting me in planning the experiments, guiding me in this thesis and supporting me in problems concerning writing and structuring this work.

Furthermore I would like to thank Prof. Dr. Christiane Wolz for giving me the possibility to do all the experiments and for giving me plenty of rope in planning and conducting these experiments, yet also for helping me when ever I needed theoretical support. I would also like to thank Vittoria Bisanzio for teaching me all the experimental skills and for helping me whenever I needed a helping hand. Moreover I would like to thank all the other people in the laboratory for their help and for the nice time.

Moreover I would like to thank all the many people that can not be mentioned here, yet who helped me directly with this thesis or just by providing me an environment and a life outside this work without which the writing of such a thesis would not have been possible. Finally I would like to thank my family (mother, father and brother) in helping me with orthography and grammar but more importantly for their support not only financially but also whenever problems concerning this thesis or the life besides this work came up.

A special thank goes to Deborah Müller for just being there, supporting me and helping me to forget all the doubts and problems and sometimes even to overcome them.

Abstract

Abstract

Staphylococcus aureus (SA) causes nosocomial infections including life threatening sepsis by multi-resistant strains *(MRSA)*. It has the ability to form biofilms to protect it from the host immune system and from anti staphylococcal drugs. Biofilm and planctonic life style is regulated by a complex *Quorum-Sensing (QS)* system with *agr* as a central regulator. To study biofilm formation and *QS* mechanisms in *SA* a Boolean network was build (94 nodes, 184 edges) including two different component systems such as *agr, sae* and *arl*. Important proteins such as *Sar, Rot* and *SigB* were included as further nodes in the model. System analysis showed there are only two stable states biofilm forming versus planctonic with clearly different subnetworks turned on. Validation according to gene expression data confirmed this. Network consistency was tested first according to previous knowledge and literature. Furthermore, the predicted node activity of different *in silico* knock-out strains agreed well with corresponding micro array experiments and data sets. Additional validation included the expression of further nodes (Northern blots) and biofilm production compared in different knock-out strains in biofilm adherence assays. The model faithfully reproduces the behaviour of *QS* signalling mutants. The integrated model allows also prediction of various other network mutations and is supported by experimental data from different strains. Furthermore, the well connected hub proteins elucidate how integration of different inputs is achieved by the *QS* network. For *in silico* as well as *in vitro* experiments it was found that the *sae*-locus is also a central modulator of biofilm production. *Sae* knock-out strains showed stronger biofilms. Wild type phenotype was rescued by *sae* complementation. To elucidate the way in which *sae* takes influence on biofilm formation the network was used and Venn-diagrams were made, revealing nodes regulated by *sae* and changed in biofilms. In these Venn-diagrams nucleases and extracellular proteins were found to be promising nodes. The network revealed *DNAse* to be of great importance. Therefore qualitatively the *DNAse* amount, produced by different *SA* mutants was measured, it was tried to dissolve biofilms with according amounts of *DNAse* and the concentration of nucleic acids, proteins and polysaccharides were measured in biofilms of different *SA* mutants.

With its thorough validation the network model provides a powerful tool to study *QS* and biofilm formation in *SA*, including successful predictions for different knock-out mutant behaviour, *QS* signalling and biofilm formation. This includes implications for the behaviour of *MRSA* strains and mutants. Key regulatory mutation combinations (*agr⁻, sae⁻, sae⁻/agr⁻, sigB⁺, sigB⁺/sae⁻*) were directly tested in the model but also in experiments. High connectivity was a good guide to identify master regulators, whose detailed behaviour was studied both *in vitro* and in the model. Together, both lines of evidence support in particular a refined regulatory role for *sae* and *agr* with involvement in biofilm repression and/or *SA* dissemination. With examination of the composition of different mutant biofilms as well as with the examination of the reaction cascade that connects *sae* to the biofilm forming ability of *SA* and also by postulating that nucleases might play an important role in this context, first steps were taken in proving and explaining regulatory links leading from *sae* to biofilms. Furthermore differences in biofilms of different mutant *SA* strains were found leading us in perspective towards a new understanding of biofilms including knowledge how to better regulate, fight and use its different properties.

Zusammenfassung

Zusammenfassung

Staphylococcus aureus (SA) ist Auslöser nosocomialer Infektionen, darunter auch die, durch multiresistente Stämme (*MRSA*) verursachte, lebensbedrohliche Sepsis. Er hat die Fähigkeit Biofilme zu bilden, um sich vor dem Immunsystem des Wirtes und vor Antibiotika zu schützen. Biofilm und planktonische Lebensweise werden durch ein komplexes *Quorum-Sensing (QS)* System mit *agr* als zentralem Regulator gesteuert. Um die Biofilm Bildung und *QS* Mechanismen in *SA* zu untersuchen, wurde ein Boole'sches Netzwerk erstellt (94 Knoten, 184 Kanten) das verschiedene Zwei-Komponenten-Systeme wie *agr, sae* und *arl* mit einschließt. Wichtige Proteine wie *Sar, Rot* und *SigB* wurden als weitere Knoten im Modell eingefügt. Die Systemanalyse zeigte, dass es nur zwei stabile Zustände gibt, Biofilm bildend versus planktonisch, in denen deutlich unterschiedliche Subnetzwerke angeschaltet sind. Überprüfungen anhand von Gen-Expressions-Daten bestätigten dies. Die Netzwerkstabilität wurde zuerst an Hand von bestehendem Wissen und Literatur getestet. Zudem stimmte die vorhergesagte Aktivität der Knoten in verschiedenen *in silico* Knock-out Stämmen sehr gut mit den zugehörigen Micro-array Experimenten und Daten überein. Zusätzliche Validierungen schlossen die Expression weiterer Knoten (Northern Blots) und die Biofilm Produktion, verglichen durch Biofilm adherence assays, in verschiedenen Knock-out Stämmen mit ein. Das Modell spiegelt zuverlässig das Verhalten von *QS*-Signal Mutanten wieder. Das integrierte Modell erlaubt auch Vorhersagen von diversen anderen Netzwerk Mutationen und wird durch experimentelle Daten unterschiedlicher Stämme gestützt. Außerdem zeigen die gut vernetzten Hubproteine im Detail auf, wie die Verarbeitung unterschiedlicher Eingangssignale durch das *QS*-Netzwerk erreicht wird. Sowohl für *in silico* als auch für *in vitro* Experimente konnte gezeigt werden, dass der *sae*-Locus auch einen zentralen Modulator der Biofilm Produktion darstellt, *sae* Knock-out Stämme zeigten stärkere Biofilme. Der Wildtyp Phänotyp wurde durch *sae* Komplementierung wiederhergestellt. Um die Art und Weise, mit der *sae* Einfluss auf die Biofilm Bildung nimmt, aufzuklären wurde das Netzwerk genutzt und Venn-Diagramme angefertigt, welche Knoten aufzeigten, die durch *sae* reguliert- und in Biofilmen verändert sind. In den Venn-Diagrammen wurden Nucleasen und extrazelluläre Proteine als vielversprechende Knoten gefunden. Das Netzwerk zeigte, dass *DNAse* von großer Bedeutung ist. Deswegen wurde qualitativ die, durch unterschiedliche *SA* Mutanten produzierte, *DNAse*-Menge gemessen, es wurde versucht den Biofilm mit vergleichbaren *DNAse*-Mengen aufzulösen und die Konzentration von Nukleinsäuren, Proteinen und Polysacchariden wurde in Biofilmen unterschiedlicher *SA* Mutanten gemessen.

Aufgrund seiner sorgfältigen Überprüfung stellt das Netzwerk-Modell ein mächtiges Werkzeug zur Untersuchung von *QS* und Biofilm Bildung in *SA* dar, erfolgreiche Vorhersagen über das Verhalten unterschiedlicher Knock-out Mutanten, *QS* Signale und Biofilm Bildung eingeschlossen. Dies beinhaltet Prognosen für das Verhalten von *MRSA* Stämmen und Mutanten. Zentrale regulatorische Mutationskombinationen (*agr$^-$, sae$^-$, sae$^-$/agr$^-$, sigB$^+$, sigB$^+$/sae$^-$*) wurden direkt im Model aber auch in Experimenten getestet. Hohe Konektivität war ein guter Anhaltspunkt, um Hauptregulatoren zu identifizieren, deren Verhalten *in vitro* und im Modell untersucht wurde. Zusammen unterstützen beide Beweisführungen im Besonderen eine präzise regulatorische Rolle von *sae* und *agr* in Bezug auf Biofilm Unterdrückung und/oder *SA* Ausbreitung. Mit der Untersuchung der Zusammensetzung von Biofilmen unterschiedlicher Mutanten, ebenso wie mit der Untersuchung der Reaktionskaskade die *sae* mit der Biofilm Bildungsfähigkeit von *SA* verbindet und auch dem Überprüfen der Annahme, dass Nukleasen eine bedeutende Rolle hierin spielen könnten, wurden erste Schritte unternommen, um regulatoische Interaktionen zwischen *sae* und *agr* zu belegen und zu untersuchen. Des Weiteren wurden Unterschiede in Biofilmen verschiedener mutierter *SA* Stämme gefunden, die uns voraussichtlich zu einem neuem Verständnis von Biofilmen und damit zu Wissen führen, wie ihre Eigenschaften reguliert, bekämpft und genutzt werden können.

Table of contents

1 - Introduction 9

1.1 - Background 9
1.1.1 - Staphylococcal physiology 9
1.1.2 - About biofilms 11
1.1.2.1 - Composition and organisation 11
1.1.2.2 - Biofilm lifecycle 12
1.1.2.3 - Live in biofilms: 13
1.1.2.4 - Biofilms in medicine 13
1.1.3 - Quorum sensing 15
1.1.4 - Synthetic biology 19
1.1.5 - About Boolean - and semiquantitative models 21

1.2 - Motivation and goal 23

2 - Material and methods 25

2.1 - Network setup and simulation 25
2.1.1 - Network setup 25
2.1.2 - Network simulation 26

2.2 - Comparative microarray analysis 27

2.3 - Bacterial strains and growth conditions 30

2.4 - Testing the mutants 31
2.4.1 - Northern blot 31
2.4.1.1 - Isolating RNA 31
2.4.1.2 - Analysing RNA 32
2.4.2 - Biofilm adherence assay 33
2.4.3 - Importance of *Sae* for biofilms 34
2.4.3.1 - Venn-diagrams 34
2.4.3.2 - DNAse concentration in biofilms 34
2.4.3.3 - Biofilm dissolution 35

2.5 - Biofilm composition 35
2.5.1 - Preparing cell-free biofilm material 35
2.5.2 - DNA detection 35
2.5.3 - Protein detection 36
2.5.4 - Polysaccharide detection 36

3 - Experiments and results 37

3.1 - Setting up the model 37

3.2 - Comparative microarray analysis (consistence of simulations with previous knowledge).. 42

3.3 - Testing the mutants (in vitro consistence with in silico predictions) 53

    3.3.1 - Northern blots 53
    3.3.2 - Biofilm strength... 56
    3.3.3 - Importance of *Sae* for biofilms 58
        3.3.3.1 - Venn-diagrams... 58
        3.3.3.2 - DNAse concentration in biofilms... 61
        3.3.3.3 - Biofilm dissolution... 64

3.4 - Biofilm composition... 68

    3.4.1 - DNA detection 68
    3.4.2 - Protein detection 69
    3.4.3 - Polysaccharide detection 70

4 - Discussion...73

4.1 - Discussing the results 73

4.2 - Related work 75

    4.2.1 - *QS* simulations around the *agr*-locus of *SA* 75
    4.2.2 - SQUAD simulation of survival and apoptosis in liver cells (Philippi et al. 2009).. 78
    4.2.3 - A *QS* regulated trade off in biofilms (Bassler et al. 2011) 79
    4.2.4 - Global gene expression in *SA* biofilms   (Beenken et al. 2004)................... 83

4.3 - Relevance of the work presented in this thesis 84

4.4 - Future work 85

    4.4.1 - Using the network 85
    4.4.2 - More detailed examination of biofilm composition. 85
    4.4.3 - Solving the DNAse dissolution problems 86

5 - Conclusion 87

6 - Appendix 89

7 - References 109

1 - Introduction

1 - Introduction

1.1 - Background

1.1.1 - Staphylococcal physiology

Staphylococcus aureus (*SA*) is a spherical, gram-positive bacterium. The old Greek word "kokkos" denotes "corn" and describes its spherical form. The word "staphyle" means "bunch of grapes" and characterises the alignment of the bacteria. The Latin word "aureus" signifies "golden" and describes the characteristic golden colour of *SA* colonies. This golden colour results from carotenoids produced by *SA*. These carotenoids act as antioxidants and are thought to protect the bacteria from noxious oxidants, like hydrogen peroxide, for example produced by the host immune system [64]. Moreover this gold pigmentation also distinguishes *SA* from other staphylococcal species. This applies to positive results of coagulase, mannitolfermentation, and deoxyribonuclease tests too [65]. *SA* has a genome that consists of a circular chromosome of approximately 2800 base pares. *SA* is aerobic or facultative aerobic, immobile, has typically a size of approximately 1μm and does not form spores [65].

SA plays an important role in many diseases, especially in nosocomial infections. However *SA* is also often part of the natural flora of many people, without causing any infections. Diseases caused by *SA* have a wide range from skin and wound infection to endocarditis and osteomyelitis and to potentially fatal systemic disorders, like the toxic shock syndrome [18]. In animals *SA* is for example one of the most important pathogenic agents in Mastitis [122]. The high virulence potential of *SA* is partly due to its *Quorum-Sensing* (*QS*) ability, as it enables *SA* to adapt perfectly to its surrounding and also to coordinate the activity of all the bacteria in one *SA* colony [see also chapter 1.1.3] [121]. *SA* for example expresses an auto inducing peptide, the *AIP*, which is the signalling peptide of a two component system composed of *agrC* and *agrA* (accessory gene regulator). *AIP* is a product of the *agr*-locus from which two RNAs, *RNAII* and *RNAIII* are transcribed, driven by two promoters *P2* and *P3*. *RNAII* codes for the proteins *agrB*, *agrD*, *agrC* and *agrA*. *AgrB* exports and processes *agrD* to *AIP* which accumulates in the *SA* colony and its concentration is proportionally rising with the density of the colony. *AIP* in turn activates *agrC* which is then able to phosphorylate *agrA* that acts as transcription factor and hence facilitates transcription of *RNAII* and *RNAIII*. With rising density of the colony this feedback loop leads to an increased *AIP* expression.

1 - Introduction

However, the also up-regulated *RNAIII* activates a complex cascade which seems -together with other factors like *SarA* (staphylococcal accessory regulator) and other two-component-systems like *Sae* (*SA* exoprotein expression)- to be important for changing the metabolic state of *SA* from a biofilm producing to an invasive, toxic phenotype [120]. A low *AIP* concentration can be found in colonies with a low bacteria density and hence the bacteria show a phenotype which is hiding from the host immune system by building a biofilm [8]. Biofilms [see also chapter 1.1.2] are protective extracellular matrices, in *SA* basically made out of Poly-*N*-acetyl-β-(1.6)-glucosamine (PNAG) -also known as polysaccharide intercellular adhesin (*PIA*)- and fibrin, which mostly comes from the host. These substances form an extracellular matrix in which the bacteria are embedded [42; 3]. When the density is rising also the *AIP* concentration is rising and the bacteria show an invasive, toxic phenotype. This phenotype disseminates from the biofilm, for example by reducing the amount of expressed surface adhesins like Staphylococcal protein A (*spa*) and fibronectin binding protein (*fnb*) and by up-regulation of the expression of proteases like *splA-F*, which dissolves parts of the biofilm and enables the cells to disseminate. Moreover *RNAIII*, besides starting a complex cascade, leading to an invasive *SA* phenotype, also is the mRNA coding for δ-haemolysin (*hld*) another important virulence factor that, together with other virulence factors like γ-haemolysins (*hlg*), which the bacteria start to produce, enable them to invade new tissue [8; 13].

Like many other bacteria *SA* can, if treated with antibiotics, gain resistance against these antibiotics. The results are multi resistant *SA* also called Methicillin-resistant *SA* (*MRSA*) strains, which are multi-antibiotic resistant and are a leading cause of hospital-acquired infections. Often Vancomycin was the last therapeutic resort, yet more and more *SA* strains acquire also resistance against this antibiotic which makes it more and more important to find new targets for the development of new anti-staphylococcal agents [48]. Such an agent could for example act by preventing *SA* from forming biofilms, when the density of the bacteria is low and hence they can easily be eliminated by the immune system. Yet such substaces could also prevent *SA* from changing from the relative harmless, biofilm forming, to the more hazardous, invasive, toxic phenotype and thus prevent *SA* from invading more healthy tissue and expressing noxious factors like the toxic shock syndrome toxin (*TSST*). Potentially all this could be reached by influencing for example the *QS* of *SA*. Yet all the variables leading to biofilm formation or the expression of virulence factors are potential targets for new anti-staphylococcal agents.

1 - Introduction

1.1.2 - About biofilms

1.1.2.1 - Composition and organisation

Most of the microorganisms living in nature don't exist planctonic, as single bacteria, yet in fact most of them live in biofilms [22]. Biofilms are layers made of different extracellular polymeric substances (*EPS*), produced by different microorganisms, organizing themselves on interfaces, mostly in aquatic systems such as the water surface or solid phases immerged in water [106]. Yet also all other interfaces for example between two fluid phases, in which microorganisms, able to form biofilms, are living can become the starting point for a biofilm. These interfaces on which biofilms develop are named the substrate. In the biofilms many different microorganisms such as other bacteria, protozoa, algae and fungi are living [106]. Yet often also only one species lives in a so called single-species biofilm. Current research focuses mainly on single-species biofilms. Examples for these single-species biofilms are *Pseudomonas aeruginosa, Pseudomonas fluorescens, Escherichia coli, Vibrio cholerae, Staphylococcus epidermidis, SA,* and *enterococci* [106].

The *EPS* produced by these microorganisms consist out of biopolymers like for example polysaccharides, proteins, lipids, and nucleic acids [62]. Yet the main component of the biofilms is water [62], together with the biopolymers it forms hydrogels giving the biofilm its stability, structure and consistency [62]. Biofilms of different microorganisms, as well as in different environmental conditions, differ a lot in their composition.

Moreover in these hydrogels bacterial substances for example for communication, attack or defence are dissolved. Yet also different exogenous particulate organic or anorganic substances like nutrients or ions, as well as different dissolved substances or gases can be embedded in these biofilms. In these biofilms also aerobe, oxygen rich regions, as well as anaerobe regions with low oxygen concentrations and within these aerobe and anaerobe microorganisms often exist only a few hundred micrometers apart [11].

We all know biofilms from the daily life, yet then we often see them as slime, plaque or coating. One popular example is dental plaque. These biofilms on the teeth can comprise hundreds of bacterial species and are subjected to a number of harsh environmental conditions, such as sparse nutrient availability, changes in pH and between aerobic and anaerobic conditions, as well as mechanic stress, every time we brush our teeth. All these influences may contribute to the regulation of biofilm development [16].

Biofilms can be deemed to be the archetype of living systems. Already the oldest fossils found about 3.2 billion years ago are showing bacteria living in biofilms. Also today the most

1 - Introduction

in vivo microorganisms are living organized in biofilms [45]. Moreover it seems obvious that these biofilms can be deemed the starting point for the development of multicellular live.

1.1.2.2 - Biofilm lifecycle

Biofilm accumulation: The accumulation of biofilms can be seen as one point in a developmental circle, with the four steps initiation, maturation, maintenance and dissolution [106]. Yet maintenance and dissolution can not be seen as steps in a chronic order but more as different steps taking place in parallel. Environmental conditions, as well as communication between the bacteria play an important role in the process of biofilm formation [56]. This communication is also known as *QS* [see also chapter 1.1.3].

Initiation: The trigger for initial attachment of the first bacteria from a biofilm varies among organisms. *Pseudomonas aeruginosa* for example forms biofilms under almost any condition, allowing bacteria to grow. Other organisms need special media or conditions such as temperature, osmolarity, pH, iron, and oxygen [106]. Some strains of *Escherichia coli* and *Vibrio cholerae* for example need complex media including amino acids [84; 117]. In contrary to that *E. coli* only forms biofilms in low nutrient media [24]. All these environmental conditions trigger a *QS* signalling cascade that leads to a change in the genetic program and thus to a change in the behaviour of the organisms. One of these changes and an important process in the initial attachment are cell-surface and cell-cell interactions. Therefore a variety of different adhesion molecules, such as polysaccharide intercellular adhesins (*PIA*) [42], are produced. By these mechanisms, organisms adhere to the interfaces and first of all form bacterial monolayers.

Maturation: When attached to a surface the bacteria change their genetic program and start producing first of all *EPS*. The bacteria become encapsulated in the *EPS*, what gives them much better environmental conditions to replicate and thus the colony grows [106]. With rising bacterial density in the biofilm, the environmental conditions, such as pH or nutrient and oxygen availability, change. This, as well as the density of the bacteria itself, can be sensed, via the *QS-* or other sensing systems, by the individual bacterium, which again leads to a change in the genetic program and the colony switches from maturation to maintenance [56].

Maintenance: The essential of this state is that the biofilm has reached the final composition and that there is equilibrium in the biofilm, regarding gain and loss of biosubstance. Changes in biosubstance could be for example due to cell division, yet also dissemination or death of bacteria plays an important role [45]. The system regulating the maintenance is again most probably *QS*. Like known from *SA* the *agr*-locus works as sensor for bacterial density in the biofilm, yet also controls genes coding for example for proteases, like *splA-F,* which dissolve parts of the biofilm and enable the cells to disseminate [8].

Dissolution: This part of the circle is probably the one with the biggest need for further investigation. Yet what we know is that the dissolution is partly due to limitations in the resources available in the biofilms. In more dense biofilms starvation and other *QS* mechanisms lead to changes in the genetic program and thus to a dispersal of the biofilm, for example by dissolving adhesive substances, like *PIA,* holding the biofilm together [42]. Another reason is, that when biofilms grow they not only form one dimensional films, yet with growing size they become multilayered and even form three dimensional structures, which are very prone to be washed away. This mechanism on the one hand limits the size of a biofilm culture, yet on the other hand these dragged away parts of the biofilm can also act as seeds for the formation of new biofilms [123].

1.1.2.3 - Live in biofilms:
Live in biofilms differs in many ways from the planktonic state which means living free floating in a suspension. A different genetic program is started due to *QS*, changes in environmental conditions and surface contact [26]. As a result for example motile bacteria loose their flagella and a lot more *EPS* are produced. This leads to a film in which the organisms are encapsulated [45; 72]. This film protects the microorganisms from the environment and increases the tolerance for example against external pH, temperature and lack of nutrition [106]. Besides that it protects the bacteria from mechanical damage and from being washed away, for example by the blood stream [26]. Moreover the organisms are better protected from the host's immune system or from antibiotics [26]. In biofilms bacteria often show reduced metabolism also known as VBNC - "viable but not culturable", as a result they show a stronger resistance to antibiotics and toxins [103]. In addition the community in biofilms benefits from synergistic effects, like when aerobia and anaerobia live together and thus the anaerobia get perfect living conditions, because the aerobia creat a perfect anaerobe milieu [11]. Besides this horizontal gene transfer becomes easier in biofilms [47].

1.1.2.4 - Biofilms in medicine
Biofilms are of great importance in medicine. Many infectious diseases are associated with biofilms. Examples are wound infections, endocarditis, or cystic fibrosis. Other than in biofilms on abiotic surfaces biofilms in medicine often also include host material such as host cells, platelets or molecules, like for example fibrin. One example for such biofilms is the bacterial endocarditis, in which in more than half of the cases *Streptococci* and in another

1 - Introduction

quarter of the cases *Staphylococci* adhere to the basement membrane. This membrane is revealed by demages in the endothelium, for example due to congenital herart defects, prostetic heart valves, vascular grafts or indwelling vascular catheters [45]. In cystic fibrosis *P.aeruginosa* is the most important germ in adolescent patients, although in early childhood most patients are colonized with *SA* or Haemophilus influenzae [58]. Most interesting in this case is that *P. aeruginosa* changes its phenotype from a non mucoid to a mucoid one, which can only be found in isolates from *CF* patients and is usually absent in environmental isolates. This mucoid phenotype showes an overproduction of the exopolysaccharide alginate and an extended resistance against antibiotic therapy. The phenotypic change is due to a deletion in the *mucA* open reading frame, most probably triggered by the host's inflammatory response to the colonization of *CF* patient's lungs [45]. There are a lot of further diseases caused by bacteria in biofilms, like for example Otitis Media. Here *SA* is the most common bacterium to form a biofilm on the mucosa of the middle ear space and thus causes severe chronic inflammation [46]. Moreover *SA* is the most common etiological agent of all septic arthritis cases in Europe. In cases where the joint has had resent injury colonization has an especially rapid progression, due to host derived extracellular matrix proteins that aid bacterial attachment, biofilm formation and thus progression of the infection. Such extracellular matrix proteins are for example fibronectin, produced to aid the defect healing in the joint. Also urinary tract infections are good examples for biofilm associated diseases. A very common germ, causing upper, as well as lower urinary tract infections, is *Staphylococcus saprophyticus*. 42% of all urinary tract infections in young women are due to *Staphylococcus saprophyticus* [2]. *SE* followed by *SA* is the most common agent causing endophtalmitis. Here planktonic bacteria cause ocular damage and infection whilst bacterial biofilms on the surface of the lenses are causing chronic infections in endophtalmitis [15]. Also of great importance are the device associated infections, which means that on medical devices like intravenous or urinary catheters, cardiac pacemakers, prosthetic heart valves, endotracheal tubes, joint prostheses, intrauterine devices, cerebrospinal fluid shunts and peritoneal dialysis catheters bacteria form biofilms. This can lead to live threatening sepsis and the device often needs to be removed [103]. Prominent examples for bacteria causing medical problems by forming such device associated biofilms are first of all the *Staphylococci*. Also of great importance are *Pseudomonas aeruginosa* or *E.coli* [103].

In biofilms bacteria are more protected against different antibiotics, although the common mechanisms like for example special mutations, modifying enzymes or efflux pumps don't play an important role [115]. Actually planktonic organisms, sensitive against some

1 - Introduction

antibiotics, might become resistant when enclosed in a biofilm. Even if the planktonic organisms remain sensitive to the antibiotics, the necessary bactericidal concentration against biofilms needed to be up to the 220 fold in the serum, then the minimum bactericidal concentration against planktonic microorganisms [54]. This means, that biofilms provide an uncommon, not that well known way of resistance, which on the other hand, when we can manage to understand the underlying principles, could also be used to develop new antibiotics and thus get more control over such biofilms. First of all there is a mechanical protection by the biofilm. Just by sealing the bacteria of from the environment and preventing the antibiotic substances from reaching them, the biofilm provides some antibiotic resistance. Moreover some antibiotics might be prevented from reaching the bacteria by electrostatic forces. One example therefore are the aminoglycosides, which are positively charged and thus bind to negatively charged polymers of the biofilm and thus cannot reach the bacteria in the biofilm [41]. Another advantageous effect of biofilms is, that only one microorganism with an active resistance, that for example produces lactamase is needed to protect all the other microorganisms enclosed in the biofilm, also because these protective substances are hold in place by the biofilm and are thus a lot more effective. Many antibiotic substances such as aminoglycosides have an optimum regarding for example pH or oxygen concentration, in which they work the best. In biofilms, due to a local accumulation of acid waste products, there are often large pH differences between the inner and outer areas; also the inner areas of many biofilms provide a more anaerobic and the outer areas a more aerobic environment. All these environmental differences in the biofilms make them a relatively safe place against antibiotics [104]. Yet the waste accumulations also have an influence on the microorganisms itself and for example cause them to enter a non-growing state, which makes them relatively resistant against antibiotics like penicillin, which target the call wall synthesis and thus only kill growing bacteria [110].

1.1.3 - Quorum sensing

QS is a common mechanism which enables bacteria to exchange signals, for example about their presence or about their metabolic state. With this mechanism bacteria can for example monitor the size of their colony, as well as different factors of their surrounding, like nutrient availability [70]. Most of the time bacteria just sense the presence of other bacteria in the surrounding. This communication is done via autoinducers, small chemical molecules resembling hormones, which the bacteria release in a certain amount to the surrounding. On their surface the bacteria express receptors, sensitive for these chemicals [116]. These receptors change the genetic program and often also up-regulate the expression of the

1 - Introduction

autoinducer itself, leading to a positive feedback [101]. Thus the concentration of these autoinducers is rising with the concentration of the bacteria, leading to a full activation of the receptor. This again leads to changes in the genetic program and as a result also to changes in the individual and also the colony behaviour of the bacteria [116]. With this mechanism microbes coordinate processes that would be inefficient or even hazardous, when doing them alone. Because this *QS* is so essential for survival and growth of different bacteria, many other prokaryotes, like bacteria competing for a special niche and also eukaryotes like for example infected hosts, try to interrupt and disturb this *QS*. This is also known as Quorum Quenching and there are already biotechnical attempts to use this, for example for the development of new antibiotic drugs [116]. Well known examples are the Bioluminiscence for example by *Vibrio fisheri* [74] as well as the formation of biofilms or the expression of pathogenicity factors, like for example in *SA* or *Pseudomonas aeruginosa* [116].

The *QS*-System of the bioluminescent marine bacterium *Vibrio fischeri* is the first described and most known *QS* system and is also considered to be the paradigm of *QS*, at least for gram negative bacteria [74]. *V. fisheri* colonizes the light organ of the Hawaiian squid *Euprymna scolopes*. In the light organ the bacterial density is rising throughout the night, leading to activation of the *QS*-system and thus to the expression of fluorescent molecules required for bioluminescence. At day the light from the light organ is used by the squid to mask its own shadow on the sea ground and thus disguising him for predation or from predators [111]. *V. fisheri* benefits from this symbiosis, by being provided with nutrients and also shelter allowing them to grow in a speed and to an amount that would never be achievable, free floating in seawater. Centre of the *QS*-System leading to the bioluminescence is the luciferase operon (*luxICDABE*), coding for molecules, important for the light production. This luciferase operon is under the control of the two genes *LuxI* and *LuxR*. *LuxI* produces the acyl-homoserine lactone (*AHL*) autoinducer which then is sensed by *LuxR* the cytoplasmic autoinducer receptor. *LuxR*, with a bound autoinducer, binds to the DNA and triggers the expression of the luciferase operon (see also figure 1-2) [116]. Autoinducers of gram negative bacteria are usually acyl-homoserine lactones (*AHL*) that are able to freely diffuse in and out of the cell across the bacterial cell membrane (see also figure 1-3) [116].

1 - Introduction

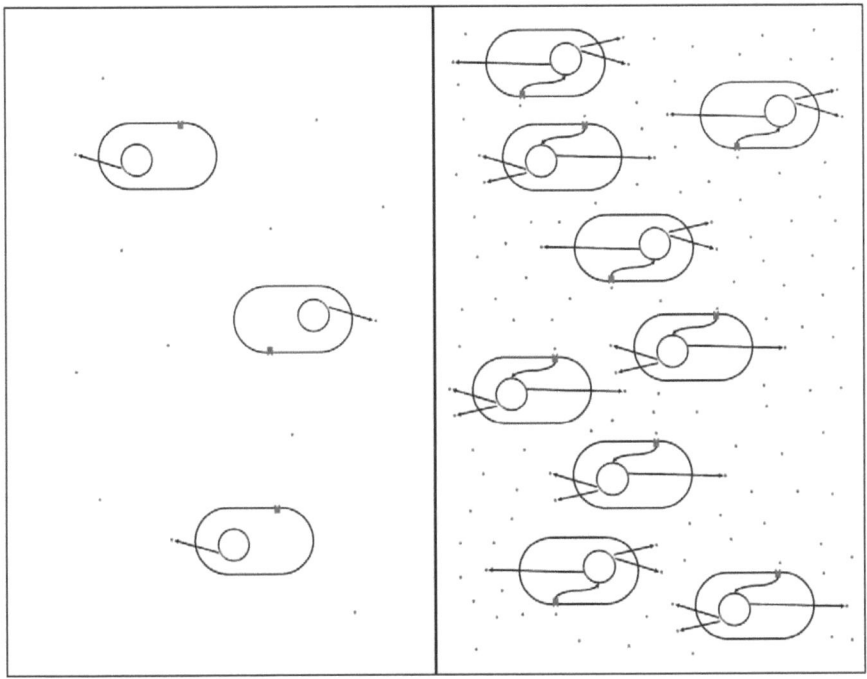

Figure 1-1: Schematic, showing the *QS* in principle: In the left part the bacterial density is low and thus there is also a low concentration of autoinducer molecules. In the right part the bacterial density is higher, thus there is a higher concentration of autoinducer molecules. The high autoinducer concentration leads to changes in the genetic program and thus to activation of further genes. The corresponding bacterial products are released. These changes often also lead to up-regulation and expression of the autoinducer itself, resulting in a positive feedback (source: own picture).

In gram positive bacteria one of the best known, most common and also very fascinating *QS*-Systems is the one around the *agr*-locus of *SA*, leading to formation of biofilms. In low cell densities the *QS*-System is inactive and *SA* produces factors, promoting attachment and colonization and also leading to biofilm production. Yet when the cell density is raising, the *QS*-System is activated, changing the metabolic state of *SA* from a biofilm producing to an invasive, toxic phenotype [120].

In contrary to autoinducers in gram negative bacteria, autoinducers in gram positive bacteria are oligopeptides, not able to cross the bacterial cell membrane without the help of specific export proteins (see also figure 1-3). Often these proteins also play an important role in processing and modification of the autoinducer. Also the receptor sensing the autoinducer can not directly bind to the DNA and thus change the genetic program. Yet the receptors are often

1 - Introduction

sensor histidin kinases, leading through a phosphorylation cascade to the activation of certain genes.

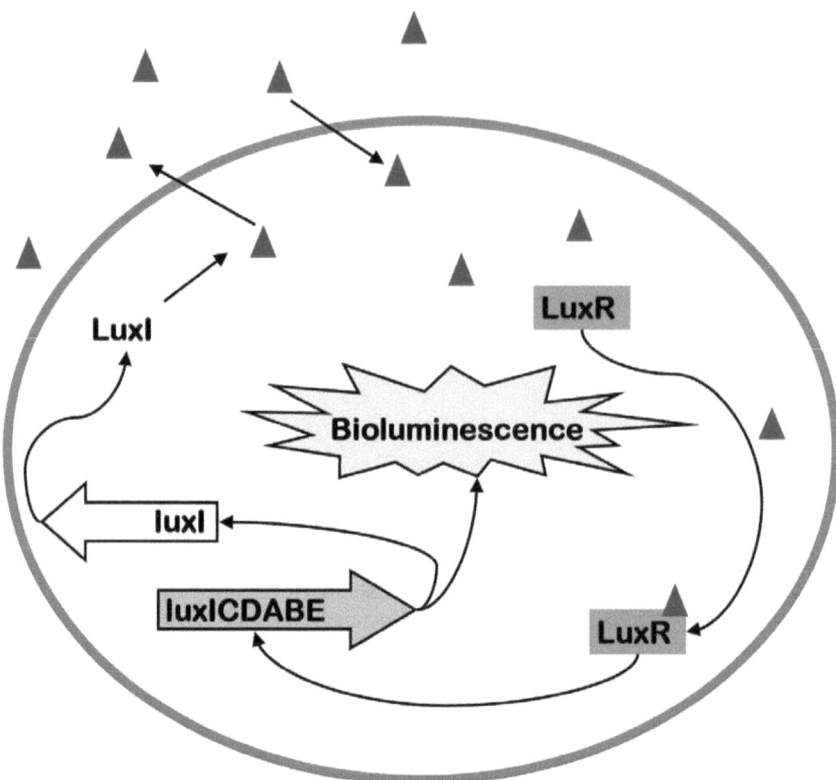

Figure 1-2: *QS* around the luciferase operon in V. fishery: *LuxI* produces the autoinducer (triangle) that can diffuse freely through the cell membrane, binds *LuxR* which then in turn activates the luciferase operon. This leads to the expression of genes important for Bioluminescence. Moreover *LuxI* is expressed leading to a positive feedback mechanism (source: own picture, modified from [116]).

The centre of this *QS*-system is the *agr*-locus with the two promoters *P2* and *P3*, coding for two mRNAs, *RNAII* and *RNAIII*. *RNAIII* represses expression of cell adhesion factors, while inducing expression of secreted factors [76] and thus amongst other regulated genes is important for the change from the adhesive, biofilm forming, to the invasive, toxic phenotype. *RNAII* on the other hand codes for the molecules *AgrA-D*, important for the *QS*-system. *AgrD* is the precursor for the autoinducing peptide *AIP*. *AgrD* is transferred through the cell

1 - Introduction

membrane by *agrB* which, in this process, adds a thiolactone ring modification to *AgrD*, to convert it into the *AIP* [92]. This *AIP* is then sensed by *AgrC* which in turn phosphorylates *AgrA*. Phosphorylated *AgrA* then binds the DNA and leads again to the expression of *RNAII* and *RNAIII* (see also fgure 1-4) [116]. Up to now four different *SA AIPs* (see also fgure 1-3) and thus four different *SA* groups are known. Each *AIP* activates its corresponding *AgrC*, yet inhibits the *AgrC* of the three other groups. As a result this system inhibits the virulence of the other groups, without affecting their growth [29].

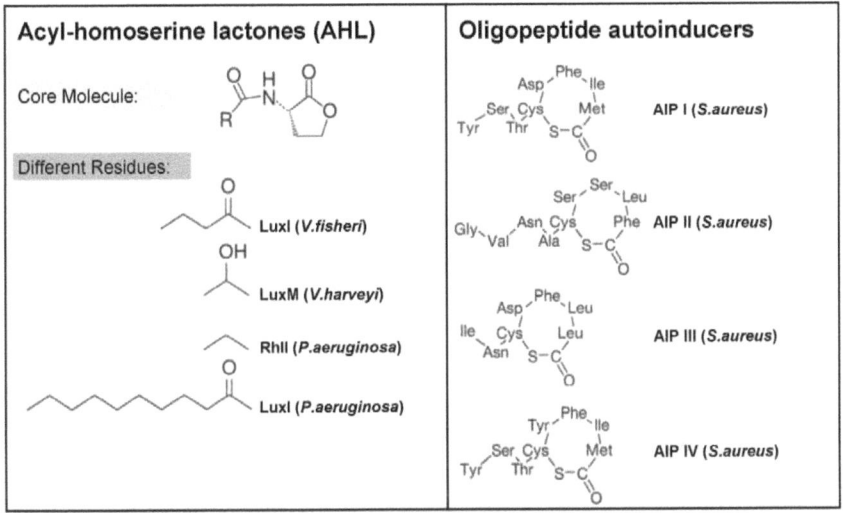

Figure 1-3: Different Autoinducers: In the left part: Examples of Acyl-homoserine lactones (*AHL*) produced by gram negative bacteria like *V. fishery*. In the right part: Examples of oligopeptide autoinducers produced by gram positive bacteria like in this case *SA* (source: own picture, modified from [116]).

1.1.4 - Synthetic biology

The field that is concerned with the investigation of how to use biological systems and structures to create, not naturally occurring, biological systems is known as synthetic biology. Biological components are independent biological parts, which can work together to function in a predefined manner. Examples therefore are biological entities such as enzymes, genetic circuits, and cells or even whole existing biological systems like for example the *QS*, metabolic pathways or signalling cascades. These biological parts are supposed to be easy to combine with each other, like Lego bricks, to make construction of artificial biological systems easier. Therefore pre-existing biological entities can be used, just as they are or be modelled and tuned to meet specific performance criteria. These entities can be compared to electronic components, such as resistors and capacitors, which are used in an electrical circuit. Thus, just like engineers nowadays design and produce integrated circuits, processors and other technical systems out of single technical entities, it is the goal of synthetic biologists to design and build engineered biological systems.

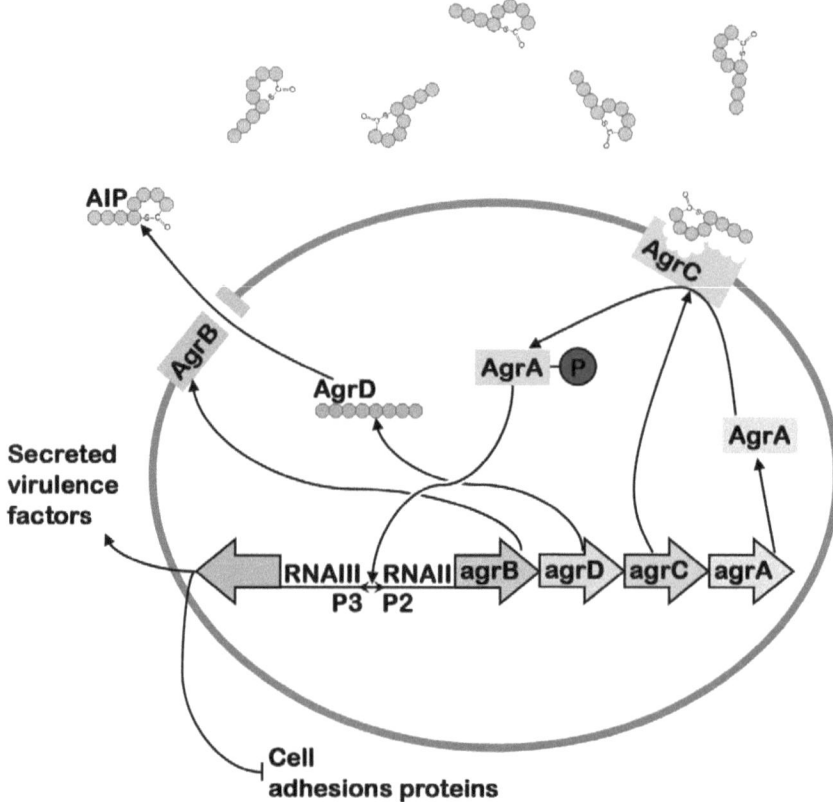

Figure 1-4: *QS* around the *agr*-locus in *SA*: The *agr*-locus comprises two promoter regions, *P2* and *P3*. From *P2 RNAII* is constantly expressed, leading to the translation of the *agr* components *AgrA-D*. *AgrD* is the precursor of the autoinducer *AIP*, in which it is transformed by *AgrB*. Simultaneous with transformation, *AgrB* transferres the autoinducer out of the cell. Here *AIP* accumulates and then binds to *AgrC*. *AgrC*, upon binding *AIP*, phosphorylates and thus activates *AgrA*. The phosphorylated *AgrA* binds the DNA and up-regulates the transcription of *RNAII* and *RNAIII*. The activation of *RNAIII* leads to down-regulation of cell adhesion proteins and to up-regulation of secreted virulence factors. The activation of *RNAII* leads to a feedback mechanism (source: own picture).

One big goal of synthetic biology is to create living systems from the scratch and then endow these systems with new and novel functions. Therefore minimal living systems, which are biological systems that get along with a minimal amount of genetic material, can be developed. These minimal systems then serve as entities that can be augmented for specific applications. The advantage of those systems is that they bring concepts such as inheritance, genetics evolution and possibly self-reproduction to artificial systems [20].

In the case of *SA* the *agr* based *QS* could be such a biological entity that can be used in new formed biological systems. This could be achieved for example just by using the *agr*-locus and with it regulating different other systems, maybe taken from other species or by constructing a new species from scratch and using the *SA QS* as biological Lego brick. Another possibility could be reducing *SA* to a minimal living system and just keeping the *agr* based *QS* and with it for example create a logic gate.

1.1.5 - About Boolean - and semiquantitative models

In the last decades there was a change in interest from just studying single molecule entities to studying their, often very complex, interaction in for example signalling cascades or other regulatory networks. Important for such networks are the connectivity (activating/inhibiting) as well as the stoichiometry and kinetic data like activation level or amount of molecules. In most cases the connectivity among the molecules is relatively easy to assess and thus available. The stoichiometry and kinetic data on the other hand are often hard to assess and thus often remain unknown. Yet for modelling a cascade or network this knowledge is essential. To close this gap a method named *Standardized Qualitative Dynamical System* was developed by Mendoza et al. [69] combining, as a hybrid modelling system, Boolean (discrete) and continuous modelling methods. Thus dynamic simulations, without knowing the exact kinetic data, became possible. In Boolean networks, nodes can only attain the two values 0 or 1, representing "inactivated" or "activated" and are connected by either positive (activating) or negative (inhibiting) relationships. The *Standardized Qualitative Dynamical System* now creates for every node in the network, by means of ordinary differential equations, a continuous range of values starting and ending with the two Boolean values 0 and 1. Boolean models are very helpful, when trying to understand complex signalling networks. With Boolean models one immediately receives semi quantitative data on all the nodes incorporated in the simulated system and also information's about qualitative changes of the whole system. Hence such Boolean models can provide helpful information's in biological or medical studies, for example when trying to understand the pathogenesis of viral infections or when searching new target sides for medicamentous interventions. One big advantage is that they are established directly from the network topology and thus work independent of detailed

1 - Introduction

kinetics. This allows simulations also with systems not explored so much in detail and can thus be a first step in elucidating principles in these systems, which then of course could be helpful again when finally revealing the actual kinetics behind the interactions. With such models the system responds to different external stimuli can be simulated and shown very easily. Also simulating modifications in the model itself such as knock out mutations is unproblematic. More than that, it is also possible to connect different Boolean models with each other given, that both networks were made following the same modelling rules. This could finally, as a long-term goal, lead to dynamic whole cell models and when combining different cell models eventually even to dynamic whole organism models. Rules that should be taken into account according to [96] are concerning the node values, how to deal with quantitativ experimental data, how to cope with time, input and output nodes, as well as unknown nodes. In classical Boolean algebra where only 1 (on) and 0 (off) exists, these two conditions are often used only to indicate the activation of a node below or above a certain level, that represents a threshold, below or above which the node has a noticeable or different effect on the network system. Yet with the possibility of continuous node simulation especially the value zero should be defined more clearly. Here it is important that the value zero should not be used for sub threshold activation anymore, yet should be used to indicate a state in which the corresponding molecule is not existing at all, like for example in a knock out mutant. On the other hand the one state should not be treated as a quantitative value, because these simulations are not quantitatively correct. Rather different quantitative values should only be taken into account when the different activation levels have different impacts on the network systems, for example when a node is activated beyond a threshold. Also quantitative experimental data can not be just transferred to the simulation, rather one needs to take into account, that only different activation levels, that are on different sides of a threshold level, should be translated to a network model. Moreover the timescale of the *in silico* simulation can not be just transferred to the *in vitro* experiments nor vice versa, because time steps *in silico* just represent the sequence of events, but not correct time steps. Thus, regarding time in the simulation, only the sequence of the events and not the exact time when they occur, can be transferred to in vitro experiments. In addition three kinds of further nodes can be implemented. First of all there are the input nodes with which a constant external input, such as environmental stimuli, can be simulated. Second there are the output nodes in which the activating and inhibiting inputs from many nodes can be collected and with which the final, overall output of the network can be simulated. One example for such an output node is the biofilm node in the network presented in this thesis. Finally there are the unknown-component-nodes, with which putative nodes, needed to explain interactions found in vitro, can be simulated and thus investigated. Yet these nodes should be marked properly as unverified and should be replaced with the actual components and interaction patterns, as soon as the underlying cellular functionality has been elucidated.

To make the next step from discreet to continuous models there are a number of different tools. Three of them are explained here briefly, regarding their potential, advantages and disadvantages. DiCara et al. for example developed a program, named SQUAD [25] that was also used for the analysis of the network model presented here. It transforms a simple Boolean network into a *Standardized Qualitative Dynamical System,* by building a system of ordinary differential equations from it, using a heuristic algorithm out of linked exponential functions, to provide a qualitative approximation of the network behavior. All equations needed for this process are generated automatically by SQUAD, upon reading a given net, xml or sbml file, in which the Boolean network is stored. SQUAD is written in Java (version 1.6.) and has an easy to handle graphical interface. The advantages of SQUAD are that it is fast and easy to use. Furthermore it immediately shows all the steady states of the examined network and allows making dynamic simulations including perturbations to identify the behaviour of the system as a whole, as well as the role and the impact of every single node. CellNetAnalyzer (CNA) [57] is a further tool to convert discreet to continuous models. It is a Matlab toolbox. The disadvantage here is that it only calculates the values of nodes, approaching a steady state and excludes oscillating nodes that sometimes exist especially in networks with feedback loops. The advantage is here that this toolbox includes a lot of tools, such as the pathway and feedback analysis tools. Another important tool is Odefy [59] a tool that uses a modelling technique called HillCube interpolation, with which it converts a Boolean model into a model of differential equations.

1.2 - Motivation and goal

As one of the most abundant nosocomial germs, often resistant to many antibiotics, *SA* plays an important role in infection biology and medicine [65]. Thus a cost efficient and easy way to test predictions about *SA* and its *QS* is a valueable tool. One way to gain a better insight, understanding and first of all overview of how the many proteins, RNAs and other molecules act together in the *QS* and thus influence the biofilm forming capability and pathogenicity of *SA* is to model the whole system. One possibility is to model it mathematically [51], which is very exact yet also very complex and not easy to work with. Applying changes in this model

1 - Introduction

is always a large effort as well. It needs a lot of time, mathematical knowledge and expertise as well as knowledge about, for example kinetic data which are often not available. Another possibility is to use a Boolean network of nodes, activating or inhibiting each other, which can easily be changed. Analysing the interaction of this network can be done very easily by using SQUAD [83]. Thus the first goal of this work is to create and provide a network to easily simulate the *agr QS* and its interactions with other important nodes, as well as its influence on biofilm forming capability and pathogenicity of *SA*. The goal was that the network can easily be modified, extended and corrected with extending knowledge and also changed for different purposes and questions. To confirm that the network is correct and properly working it was first tested against previous knowledge about the *QS* and global gene regulation networks in *SA* colonies, by conducting a comparative microarray analysis.

Secondly, concerning the *QS*, gene regulation and biofilm building ability in *SA, in silico* predictions were compared, by Northern blot and biofilm adherence assays, with the real world, more exactly with *in vitro* results. Moreover the influence of *DNAse* on the *SA* biofilm was tested and its composition was analysed.

With this model a validated tool is presented to analyse *QS* and genetic or pharmacological modifications. It has applications in pharmacology and medicine as well as for biology and basic research, for instance regarding the cellular differentiation switch from biofilm forming to planctonic life style of gram positive bacteria. Furthermore, it allows examining network modification for *QS* signalling and testing new potential pharmacological interventions against biofilm formation to prevent severe infections with Staphylococci. Yet this network is not ment to be the final version, more than that it should represent a comprehensive network reflecting the resent state of knowledge and thus a basement for easy examining newly found nodes and substances like non coding mRNAs. Such new, potentially regulatory components are always promising targets for the development of new anti staphylococcal substances and there importance could easily be examined using the network presented here.

2 - Material and methods

This chapter explains all the experiments done for this thesis. All the bioinformatical analyses and also all the experiments described here, where no references are mentioned, were done by myself. The only experiments not done by myself are the actual microarray experiments, which were retrieved from SAMMD database, as well as the construction of the knock out *SA* strains which where kindly provided by Prof. Dr. Christiane Wolz and her group. Moreover selecting and planning the experiments was in large parts my work, of course always with a helping hand and the support of my two supervisors Prof Dr. Thomas Dandekar and Prof. Dr. Christiane Wolz.

2.1 - Network setup and simulation

2.1.1 - Network setup

For setting up the network, information about different nodes and their interaction were collected from different databases such as KEGG (http://www.genome.jp/kegg/) and STRING (http://string.embl.de) and a first basic network model was created. Moreover, an extensive literature research was done, first to collect further information about all the nodes and include further components in the network. Important for the construction of this network model was, that it was aimed to get a detailed comprehensive network that includes all the important nodes, yet only on a level taking the kind of interactions into account and not in which way they come about. Thus details such as whether the activation or inhibition influences the transcription, the translation or the efficiency of the final gene product were not included. However in some cases the interaction is due to protein-protein interactions and in some cases of course also due to protein or RNA gene interaction as known for example from transcription factors. Yet the mechanisms of interactions are not that important for this network model and are also often not revealed until know. In a lot of cases all we now for example from knock-out experiments is that there is an inhibiting or activating influence from one node on another one. The main work in modelling this network, besides combining the above mentioned sources was the simulation itself. This was done by directly regulating the activity of different nodes with the aid of SQUAD to simulate different scenarios. The detailed properties of these simulations are described in the corresponding sections. Finally a comprehensive network system was achieved, in which all the nodes that are known to be of importance for the *QS* process around the *agr*-locus, as well as all the known activating and

2 - Material and Methods

inhibiting connections between these nodes, are included. The construction and visualization of the network was done by using a computer program named Cell-designer Version 3.5.1 [35] (www.celldesigner.org). With this program interacting networks can be created, just by creating a node and connecting it with either inhibiting or activating links to other nodes. The result is a Boolean network with the Boolean states (either 0 = off or 1 = on) for each node. This output of Cell-designer can be stored in the SBML format.

2.1.2 - Network simulation

The output of CellDesigner was further analysed, using SQUAD Version 2.0 [25], to identify and calculate the amount of stable steady states for the node-node interaction network. There is a high number of possible states in a Boolean network: In the network presented here, with 94 nodes there are 2^{94} possible states considering that each node can just be either active or inactive. However, in most cases the system states are unstable, as the network logic rapidly transforms the state into the next state. Only very few system states remain stable and thus the activation of their nodes remain in equilibrium. These are the stable states for the node-node network and evolution and selection made sure that these correspond to clear biological functions.

To calculate these steady states, one first has to consider that the more precise the simulation should be the more increments between 0 and 1 are needed and thus the amount of possible network states tends to become infinite. From this large amount of system states the interesting ones need to be selected. The interesting states are generally the stable ones, to which the network aspires and to whom it always comes back after a temporally limited external input and in which the expression of each node is stable.

When analyzing the existence of steady states in the networks, SQUAD converts the Boolean models into a dynamic system in which the Boolean states (either 0 = off or 1 = on) are replaced by a heuristic curve (catenated e-functions according to network topology, also see [16]) to interpolate all possible intermediate states of each node. For this process SQUAD, in a first step, uses a fast heuristic search algorithm to identify the existing stable and steady states in the discrete node-node interaction network. In a second step the steady states in the dynamical continuous model are calculated, with activity states ranging from 0 to 1, for each node. All the equations necessary for this calculation process are generated automatically by SQUAD, when processing the loaded file. With this algorithms SQUAD is able to identify steady states also in complex networks, even when kinetic parameters are not available. In the network model presented here, the SQUAD simulation found two steady states, which

2 – Material and Methods

correspond well (regarding resulting behaviour of the whole network, as well as the activation level of individual nodes) to two biological states (biofilm and planctonic), that are actually observed in *SA*.

SQUAD includes dynamic simulations where external parameters can be changed in an ongoing simulation. The initial activation strength of each node is selected and whith this also the influence each node has, with its inhibiting or activating interactions, on the other nodes and on the whole system. The strength of the, either activating or inhibiting, interactions corresponds with the activation level of the nodes of their origin. The activation level of all the nodes in the system will then be regulated in course of the ongoing simulation, according to their, either activating or inhibiting, inputs. In the end a state is achieved, in which the activating and inhibiting signals to each node are in equilibrium and thus the activation level of all the nodes in the system is not changing anymore. This state is then called the steady state. It obviously depends on the initial conditions which of the existing steady states the system acquires finally. Also external inputs can transfer the system state from one existing steady state to another one. This is achieved by changing the balance of the node activation level to a pattern that leads, with proceeding simulation, to a balance in the noden activation level, representing one of the other steady states, existing in the network.

Moreover, the activation level of few nodes can also be fixed independent from the simulated input signals to these nodes. This can be used for example to represent a knock out strain or any other stable external stimulus, applied on the system. This possibly changes the system as a whole and maybe in consequence also changes the amount of steady states, as well as their node activation distribution. In this way the knock out mutants, as well as a stable external milieu were simulated in the in silico experiments.

All simulations were done on a computer with Windows Vista Home Premium SP2 (32Bit), 4GB RAM and an Intel core 2 duo CPU with 2.53GHz.

2.2 - Comparative microarray analysis

In this bioinformatical analysis the *in vitro* changes in gene expression, due to genetic or environmental changes, were compared to corresponding *in silico* node activity changes. For this analysis data from microarrays of different *in vitro* scenarios retrieved from the SAMMD database (http://www.bioinformatics.org/sammd/) was used. The concordance was investigated between *in silico* and *in vitro* in three different scenarios. **A**: agr^+ vs. agr^-; **B**: $sarA^+$ vs. $sarA^-$ and **C**: $biofilm^+$ vs. $biofilm^-$. In the scenarios **A** and **B** all nodes, showing in

2 – Material and Methods

the wild type strain a robust activation level, at least three times higher than that in the mutant strain, were regarded as up-regulated. On the other hand, it was regarded as down-regulation of the gene, when its expression in the wild type strain was at least three times lower than in the mutant strain. In scenario **C** a gene was regarded as up-regulated, when its expression was 2.5 times higher in the biofilm forming situation than in the biofilm negative one. On the other hand a gene was declared as down-regulated in the biofilm forming situation when it was at least 2.5 times stronger expressed in the biofilm negative situation, compared to the biofilm forming situation.

In SAMMD data was gathered from different experiments that fit the three scenarios. For scenario **A** and **B** microarray analyses conducted by Cassat et al. [17] with UAMS-1 *SA* strains were used. This analysis compares the microarray gene expression data of agr^+ strains to the expression data of agr^- strains. This was done in the exponential phase (OD 1.0 at 560nm) and in the post exponential phase (OD 3.0 at 560nm). For this microarray analysis in the same way $sarA^+$ and $sarA^-$ strains at OD 1 and OD 3 were compared. So finally four different datasets were acquired (**A1**: agr^+ vs. agr^- *OD1*; **A2**: agr^+ vs. agr^- *OD3*; **B1**: $sarA^+$ vs. $SarA^-$ *OD1*; **B2**: $sarA^+$ vs. $SarA^-$ *OD3*). Moreover microarray analyses were used, conducted by Dunman et al. [30] with RN27 *SA* strains. This analysis compares the microarray gene expression data of agr^+ strains to the expression data of agr^- strains and the expression data of $sarA^+$ strains to $sarA^-$ strains at OD 3. So two more datasets were acquired from this analysis (**A3**: agr^+ vs. agr^- *RN27*; **B3**: $sarA^+$ vs. $SarA^-$ *RN27*).

For scenario **C** three different datasets were collected from SAMMD. For dataset **C1** microarray analyses conducted by Brady et al. [12] was used, where the gene expression of a late exponential phase (6h) planktonic culture is compared to the gene expression of a maturing (48h) biofilm culture. For dataset **C2** microarray analyses conducted by Resch et al. [86] was used, who compared the gene expression of an, in biofilm grown (for 24hrs) *SA113* colony, with the gene expression of a planktonically grown (for 24hrs) *SA113* colony. For dataset **C3** microarray analyses conducted by Beenken et al. [6] was used. In this analysis the gene expression of a one week old biofilm forming *UAMS-1* colony, grown in flow cell, was compared to the gene expression of a planktonically grown *UAMS-1* colony in the stationary phase (OD 3.5 at 560nm).

Finally all these *in vitro* datasets were compared to corresponding *in silico* datasets. The datasets from the **A** and **B** scenarios were compared to datasets, created by comparing the node activation levels in SQUAD simulations, under wild type strain settings, to node activation levels in the SQUAD simulations of *sarA⁻* and *agr⁻* strains under T1 and T3 circumstances respectively. For this the same settings were used as for the Northern blot simulations. Here four different datasets were acquired: *agr* sim T1; *agr* sim T3; *sarA* sim T1 and *sarA* sim T3. Each dataset from scenario **C** was compared to two different *in silico* datasets, obtained from SQUAD simulations. For creating the first *in silico* dataset just the activation level of the nodes in the biofilm producing steady state 2 was compared to the biofilm negative steady state 1 (**sim1:** SS2 vs. SS1). For creating the second *in silico* dataset just the *AIP* concentration was changed by up-regulating the activation level of the *AIP* node to simulate the influence of the *QS* processes on the biofilm formation ability. For this the activation levels of the nodes under biofilm faciliating circumstances, due to low activation levels of the *AIP* nodes, were compared with the node activation levels under biofilm repressing, high *AIP* concentration, settings (**sim2:** *AIP* low vs. *AIP* high).

For further analysing the reactions of the network model and the analogy of the model in layout and reaction to experimental and already published data and to make the results more comprehensible some further graphics were prepared. First of all to see in which way which node is affected, when *agr* or *saeRS* is knocked out *in silico* and also to see if these reactions agree with what is known from already published experiments about the behaviour of the different nodes, figures were prepaed in which the activation level of the nodes in *agr⁻* was compared to that in *agr⁺* and also the activation level of the nodes in *saeRS⁻* was compared to that in *saeRS⁺* (see figures 3-3 and 3-5). In the figures, nodes that are down-regulated are marked in dark gray and nodes that are up-regulated, in the mutant strain, are marked in light gray. To underline that these results are already described in literature and can not only be found in the *in silico* experiments, Additionally highlighted in these figures are all the names of the nodes, from whom it is known out of the literature that they show this reaction under the same conditions. To show in which supporting reference this node behaviour can be found little flags in different shading were added to the nodes, indicating the different references.

Furthermore, to get a more detailed insight in how prominent nodes (*ArlR, hla, icaA-C, RNA III, Rot, SaeR, SarA, sigB* and *sspA*) are affected in *agr* or *saeRS* knockout strains, graphics were created, showing the *in silico* activation change of these different prominent nodes, as a result of knocking out *agr* (see figure 3-4) or *saeRS* (see figure 3-6). In these graphics nodes, down-regulated under knock out conditions, were highlighted by writing them in light gray. To show in which reference a corresponding node behaviour under these conditions is already described little flags in different shading were added to the nodes, indicating the different references.

2 – Material and Methods

A change between wild type and mutant was assumed *in silico* and *in vitro*, when the expression strength in the mutant was 2.5 fold stronger or weaker than in the wild type. In all in silico scenarios the different two component systems (TCS) incorporated in the simulation were up-regulated a little bit to simulate an *in vitro* like surrounding, where the TCS are putatively stimulated to some extent.

2.3 - Bacterial strains and growth conditions

The strains used for the different experiments are listed in Table 2-1. Strain ISP546-29 was obtained by transduction using a φ11 phage lysate of strain ISP546 and strain ISP479C-29 as recipient.

Bacteria were taken from the frozen stock, provided by AG Wolz and streaked out on blood agar plates, filled with tryptic soy agar [Oxoid; Basingstoke; UK] and with the appropriate antibiotics. On these blood agar plates bacteria were grown for approximately four days at 37°C. For getting an overnight culture the bacteria were picked from the blood agar plates and transferred to test tubes, each filled with 5ml CYPG culture medium. To each test tube 5µl of the antibiotic was added, against the strain was resistant (Erythromycin (10 mg/ml) [BioChemica; Buchs; Switzerland] Strain Nr. 4; Kanamycin (50 mg/ml) [AppliChem; Darmstadt; Germany] Strain Nr. 3, 3b, 4, 6). Only Tetracycline (5 mg/ml) [Serva Feinbiochemica; Heidelberg; Germany] (Strain Nr. 2, 3b) was used in a lower dose of 2.5µl per test tube. The bacteria were then grown, again for one night, constantly shaken, at 37°C. After one night the culture was diluted 1:100 with CYPG and the optical density (OD) was measured at 600nm with a Photometer. Day cultures were prepared in new test tubes, each filled with 12ml CYPG. The bacteria were inoculated with an OD at 600 nm (OD600) of 0.05 in CYPG and again incubated at 37°C in the shaker. The bacteria were allowed to grow to the mid-exponential phase, (OD of 0.5 (T1)), approximately 2.5h after inoculation. Afterwards 7ml of the probes were harvested. The rest of the cultures were again allowed to grow until they reached an OD of 1.5 (T2). This time no bacteria were taken out of the culture, yet they were grow for further 1.5h (T3) to the post exponential level, approximately 7h after inoculation and than harvested 1ml of each culture.

2 - Material and Methods

Table 2-1: Bacterial strains:

Strain Nr. / Type	Strain	Description	Phenotype	Reference
1. WT	ISP479C	Parental strain, subculture of strain 8325-4, Sigma factor B deficient	rsbU defect, agr^+, sae^+	82
2. agr^-	ALC14	$\Delta agr::tetM$	rsbU defect, agr^-, sae^+	118
3. sae^-	ISP479C-29	$\Delta saePQRS::kan$	rsbU defect, agr^+, sae^-	66
3b. sae^- rescue	ISP479C-29, pCWSAE28	$\Delta saePQRS::kan$, with integration plasmid pCWSAE28 containing saePQRS for complementation	rsbU defect, agr^+, sae^+	66
4. sae^-/agr^-	ISP546-29	$agr::Tn551$, $\Delta saePQRS::kan$	rsbU defect, agr^-, sae^-	This study
5. $sigB^+$	ISP479R	ISP479C, in which the mutation in rsbU was repaired to gain full sigma factor B activity	rsbU repaired, agr^+, sae^+	105
6. $sigB^+/sae^-$	ISP479R-29	restored rsbU, $\Delta saePQRS::kan$	rsbU repaired, agr^+, sae^-	36

This table shows the *SA* strains, used for Northern blot analysis, biofilm formation and biofilm investigation experiments.

2.4 - Testing the mutants

2.4.1 - Northern blot

2.4.1.1 - Isolating RNA

The bacteria were centrifuged at 5000g, for 5 minutes, at 4°C. The supernatant was discarded and the *SA* cells were lysed by suspending the pellet in 1 ml Trizol (Invitrogen/Life Technologies; Grand Island, NY; USA). The suspensions were then transferred to 1.5ml cups, filled with 0.5 ml Zirconia-Silica beads (diameter, 0.1 mm) (Carl Roth; Karlsruhe; Germany) and two times shaken for 20 sec. at 6500rpm, in a high-speed homogenizer (Fastprep; MP Biomedicals; Irvine, California; USA). Thereafter 200µl chloroform was added to the suspension, shaken manually for one minute and incubated for further three minutes. Afterwards the probes were centrifuged at 12.000g, for 15min, at 4°C. The pellet was discarded and the supernatant was transferred to new 1.5ml cups, filled with 500µl Isopropanol and incubated for ten minutes at room temperature. Afterwards the samples were again centrifuged at 12.000g, for 30min, at 4°C. This time again the supernatant was

2 – Material and Methods

discarded and the pellet was washed with 500µl 70% ethanol. The RNA was again centrifuged at 7500g, for 5min, at 4°C and the supernatant discarded. The pellets were dried for 60min, just by leaving the lid of the cup open. After the pellet was dried and became vitreous 50µl 1mM natriumcitrat was added and everything was incubated for 10min at 55°C. Then the probes were vortexed for 4min and stored on ice.

Now 2µl of the samples were diluted 1:200 with RNA-Water and the amount of RNA in the samples was determined using the photometer again.

2.4.1.2 - Analysing RNA

Preparation of blotting probes: Based on the amount of RNA the needed volume for each solution was calculated and the samples were diluted with nuclease free water [Ambion/Life Technologies; Grand Island, NY; USA], to get solutions containing 8µg RNA each in a volume of 8µl. Hence the solution was used for four gels; 2µg RNA of each probe were used for each gel. 24µl probe buffer and 6µl Blue Juice [Invitrogen/LifeTechnologies; Carlsbad; CA, USA] was added to the RNA. Then the solutions were incubated for 15min, at 65°C, on a heating block [QBD2; Grant Instruments; Cambridge; UK]. Finally 9.2µl of each probe was transferred into one pocked of each gel.

Preparation of the gel: RNA was separated on a Formaldehyde-Gel (a 1% agarose-0.66M Formaldehyde-Gel). Therefore 0.8g Agarose [Biozym Scientific; Hessisch Oldendorf; Germany] was melted in 68ml RNA free water. After cooling to approximately 45°C 8ml 10x MOPS [AppliChem; Darmstadt; Germany] and 4ml Formaldehyde [AppliChem; Darmstadt; Germany] was added. Before being completely cooled down the gel was poured into prepared containers, with combs to form slots for the RNA samples. After formation of solid gels the combs were removed. The gels were then put into 1x MOPS and the RNA samples were applied in slots of the gels. Finally 65 volts were applied, for approximately 4h, to move the RNA through the gels.

Blotting: After electrophoresis the intensities of the 23S and 16S rRNA bands, stained by ethidium bromide, were checked to be equivalent in all the samples. Therefore the gels were photographed. Then the gels were three times washed for ten minutes with RNA free water. The gel was blotted by alkaline transfer (Turbo Blotter; Schleicher and Schuell; Dassel; Germany). The gel was placed onto a positively charged nylon membrane (Biodyne; Pall Corporation; Port Washington, New York; USA) soaked in transfer buffer. Above and below blotting papers were placed, also soaked in transfer buffer. Below this package further dry blotting papers were placed. Above al this, a long blotting paper was placed, soaked in

2 – Material and Methods

transfer buffer too, with both ends immersed in transfer buffer. Everything was covered with cellophane foil and an approximately 500g wait was placed on top of it. After 2.5h all the blotting papers were removed and the pockets of the gel, as well as the location of the RNA were marked on the nylon membrane with a pencil. Afterwards the nylon membrane was washed for 5min in 1x phosphate buffer and cross linked from each side with UV light.

The blot was hybridized with digoxigenin-labeled RNA probes, following the instructions given by the manufacturer of the DIG wash and block buffer set (Roche; Mannheim; Germany). Therefore the nylon membrane was transferred to a glass tube, with the RNA side facing the centre of the tube. These tubes were then rotated at 64°C in an oven, to make sure the whole membrane gets contact to the different chemicals then applied to the tubes. First 15ml hybridisation buffer "High SDS" was applied and incubated for 30min. Afterwards 5µl hybridization probe diluted 1:2000 to detect specific RNA sequences were added, after they had been boiled for ten minutes in a water bath. In this case hybridization probes against *asp* (only up-regulated by *sigB* and hence in its activity nicely related to the activity of *sigB*), *sarA*, *agr* (*RNA II*) and *sae* were used. The hybridization was performed over night at 64°C. On the next day the nylon membrane was incubated first two times for 5min with 15ml of 2x SSC/0.1% SDS and second two times for 15min with 15ml of 0.2x SSC/0.1% SDS. The following incubations were done at room temperature. The first incubation was done for one minute with 15ml washing buffer, followed by 30min incubation with 15ml of a blocking solution. Now 10 ml antibodies recognising digoxigenin were applied (1:10.000 dilution in blocking solution). Then the membrane was washed two times for 15min with 15ml washing buffer, afterwards incubated for two minutes with 2ml detection buffer and finally for 5min incubated with 2ml CSPD [Roche; Mannheim; Germany] in a 1:100 dilution with detection buffer.

The nylon membrane was then removed from the glass tubes and the Bioluminescence signals of the CSPD bound to the antibodies were detected with a film (Agfa Curix HT1000 G Plus Folienfilm; Agfa; Mortsel; Belgium). Therefore the membranes were wrapped in cellophane foil, placed on the film and incubated for 2.5h at 37°C. Finally the film was developed and scanned to the computer.

2.4.2 - Biofilm adherence assay

For detecting the strength of the biofilm forming ability of the different knock-out mutants a biofilm adherence assay was conducted, similar as described by Christensen et al. [21]: Bacteria were grown, as described above for the Northern blot, but in TSB (tryptic soy broth)

2 – Material and Methods

(Oxoid; Basingstoke; UK) with 0.5% glucose instead of CYPG. T1 bacteria were then collected and again diluted to an OD of 0.05. 1000µl were then transferred to each well of a 24-well plate (Greiner Bio-One; Frickenhausen; Germany). These 24-well plates were incubated for 24h at 37°C and washed three times with 1ml PBS (phosphate buffer saline). Afterwards the biofilm was fixated with 1ml of 50% methanol for approximately 30 minutes. In a next step the biofilm was dyed with 200 µl Carbolgentianviolett for one minute. Excessive dye was removed, by three times washing with water. Afterwards a photo was taken from the 24-well plate showing in blue colour the biofilm adhering to the bottom of the different wells. Whether a strain was biofilm forming or not was evaluated simply by comparing the intensity of the blue colour without technical support [21], because there was an obvious visual difference in the biofilm intensity between biofilm positive and biofilm negative strains. Moreover, according to Christensen et al. [21], evaluating with or without technical support, whether a strain was biofilm positive or biofilm negative, leads to the same results.

2.4.3 - Importance of *Sae* for biofilms

2.4.3.1 - Venn-diagrams

For finding the gene, responsible for the biofilm inhibiting effect of *sae* in *SA* the microarray gene expression data of five sae^- vs. sae^+ experiments [89; 60] and ten in biofilm vs. Planctonic grown strains [12; 86; 6] was used. All the five sae^- vs. sae^+ experiments were compared to the ten biofilm vs. planctonic experiments and Venn-diagrams of these fifty combinations were made.

2.4.3.2 - DNAse concentration in biofilms

The strains that were tested for their *DNAse* production were: 1. a *wild type* strain, 2. an agr^- mutant, 3. a sae^- strain and 4. an agr^-/sae^- double mutant strain. These strains were grown as described for the biofilm assay. Yet instead of inoculating them in 24 well plates especially prepared agar plates were used. These plates are made out of tryptic soy agar with DNA. In these plates little holes were made, approximately 2-3mm in diameter, by using heated metal inoculation loops. In each of these holes 10µl of one of the *SA* strain cultures were transferred. After growing the strains in these holes overnight 1N HCl was poured on the plates. In the regions where DNA was left the agar turned gray and intransparent. In the surrounding of the strains transparent areolas were formed, differing in their size, according to the amount of *DNAse* produced by the strain, inoculated in the coresponding hole. Moreover to get an idea of approximately how much *DNAse* was produced by the different strains, the same experiments were made with *DNAse* solutions of different concentrations (50U/ml, 100U/ml, 200U/ml, 300U/ml and 400U/ml).

2.4.3.3 - Biofilm dissolution

For these experiments the biofilm adherence assay was used as described above. Yet after growing the biofilms overnight they were only washed two times with 1ml PBS (phosphate buffer saline) and then inoculated for five hours with 200µl of *DNAse* [AppliChem; Darmstadt; Germany] in different concentrations (see table 3-6. and figure. 3-14). Because of the astonishing results obtained from these tests, the decision was made to do further experiments in which the effect of different components of the *DNAse* solution on the biofilm was checked (table 3-6. and figure. 3-15). After the five hours of inoculation at 37° the biofilms were again washed two times with 1ml PBS (phosphate buffer saline) and then the assay was completed, as described above in the "biofilm adherence assay" section (see chapter 2.4.2).

2.5 - *Biofilm composition*

2.5.1 - Preparing cell-free biofilm material

The strains of which a cell free solution was prepared for further analysis of the biofilm composition were: 1. an *agr*⁻ mutant, 2. a *sae*⁻ strain and 3. an *agr*⁻/*sae*⁻ double mutant strain. For these experiments, again the biofilm adherence assay was used, as described above. Yet instead of growing the biofilms in 24 well plates cell culture flasks were used (Cell Culture Flask 75cm^2/250ml; Greiner Bio-One; Frickenhausen; Germany) in each of which 45ml of the liquid *SA* culture was poured. After growing the biofilm for approximately 18h the extensive liquid culture was poured out and the biofilms were washed two times with 25ml PBS. Afterwards the biofilm was dissolved in 10ml NaCl/EDTA (9.375ml 0.14 mol/l NaCl + 0.625ml 0.5 mol/l EDTA). This solution was centrifuged for 3min at 5000rpm and filtered through a sterile filter with a mesh size of 0.22µm. Immediately thereafter the filtered biofilm-solution was frozen and if needed 1.5ml aliquots were carefully melted and used for the following experiments. Such biofilm-solutions were produced three times from each strain, thus nine different samples were acquired.

2.5.2 - DNA detection

Qualitatively quantifying the amount of nucleic acids in the different biofilms was done by blotting (see figure 3-16). Therefore 48µl of biofilm-solution from the different strains were

2 – Material and Methods

loaded onto an agarose gel, made out of 300ml TAE with 4.5g agarose (Biozym Scientific; Hessisch Oldendorf; Germany) and 30µl Gel Red (Biotium, Inc; Hayward, CA; USA). The biofilm solutions were electrophoresed for approximately 20 min at 140V. Thereafter a photo was taken from the gels. The amount of nucleic acids in each of the nine probes was quantified four times, thus 36 results; 12 results for each strain were acquired.

2.5.3 - Protein detection

Qualitatively quantifying the amount of protein in the different biofilms was done by a Bradford-Protein-assay (see Table 3-7) [10]. Therefore a Bradford-Solution was prepared out of 60mg Coomassie Brilliant Blue G-250 (Sigma-Aldrich; St. Louis, Missouri; USA) and 1l 3% Perchlorsäure (Merck; Darmstadt; Germany). 500µl biofilm-solutions, with 500µl Bradford-solution were measured within 2min-60min, at 595nm, against 500µl H_2O, with 500µl Bradford-solution. The amount of protein in each of the nine probes was measured four times. The mean of these four measurements was calculated and the strains were ranked by the amount of protein in their biofilms.

2.5.4 - Polysaccharide detection

Qualitatively quantifying the amount of polysaccharides in the different biofilms was done by a Polysaccharide-assay (see Table 3-8), like described by Dubois et al. [27]. Therefore 200µl 5% Phenol-solution was given to 200µl of the biofilm-slutions. Within 5sec-10sec, 1ml concentrated H_2SO_4 was added and the probe was vortexed. The probes were left at room temperature for 10min. Thereafter the probes were kept at 30°C for 15min and then again at room temperature for 5min. The OD of the probes at 480nm was measured against H_2O, prepared for measurement with Phenol-Solution and H_2SO_4. The amount of polysaccharides in each of the nine probes was measured four times. Each time the probes were prepared again for measurement with Phenol-solution and H_2SO_4. The mean of these four measurements was calculated and the strains were ranked by the amount of polysaccharides in their biofilms.

3 - Experiments and results

3.1 - Setting up the model

For setting up the network different two component systems (TCS) were selected, supposed to be important in *SA* for detecting the colony density and the environmental conditions (e.g.: *agr, sae*) [13; 77; 89], as well as different nodes, also supposed to be important for the quorum and environment sensing network (e.g.: *sar; sigB*) [13, 77]. Then data was collected on these nodes, using databases like STRING, as well as gen expression data and already existing network models from different publications (for a complete summary of all references see Table: 3-2). This data was then compiled into one complete network with 94 nodes and 184 edges, using Cell-designer v.3.5.1. The network represents different TCS and also signalling cascades that connect these TCS and lead to either up-regulation or depression of the biofilm forming capability of the *SA* colony. This network is thus an overview of the knowledge we have today about the *agr*-locus and its signalling cascade, influencing different important nodes in regulatory vicinity of it. If the current knowledge about the *agr*-locus and its signalling cascade is correct, the network reactions should be consistent with the knowledge we have until now about alterations in the signalling cascade and resulting changes of the phenotypic output, depending on different changes of nodes in the network. Moreover this network should be able to be used for predictions about alterations in the network signalling cascade, as well as for predictions about changes in the phenotypic reactions of *SA* colonies, again dependent on changes of different nodes in the network.

Table 3-1: Activity of different nodes in the two steady states (SS1 and SS2):

Node	Activation in SS1	Activation in SS2
abcA	1.000000000	1.000000000
agr	0.929121724	0.000000000
AgrA-P	0.929896264	0.000000000
AgrB	0.999898610	0.000000000
AgrC	0.999999923	0.000000000
AgrD	0.999898610	0.000000000
AIP	0.999997722	0.000000000
aur	0.999567678	0.070103715
biofilm	0.000103676	0.913598828
clfB	0.000582424	1.000000000
coa	0.004164738	0.000000000
DNAse	0.999999923	0.000000000
eap	0.845758097	0.000000000
emp	0.845758097	0.000000000

3 – Experiments and results

fnbA	0.000008953	0.000000000
geh	0.999417576	0.000000000
hla	0.981777638	0.000000000
hlb	0.998347767	0.000000000
hld	0.995835256	0.000000000
hlgA	0.999999923	0.000000000
hlgB	0.925356954	0.000000000
hlgC	0.925356954	0.000000000
icaR	1.000000000	1.000000000
lytN	1.000000000	1.000000000
norA	1.000000000	1.000000000
norC	1.000000000	1.000000000
RNA_II	0.997017139	0.000000000
RNA_III	0.907882820	0.000000000
Rot	0.004164744	1.000000000
RsbW	1.000000000	1.000000000
SaeP	0.999430482	0.000000000
SaeQ	0.999430482	0.000000000
SaeR	0.999898610	0.000000000
SaeR-P	0.999997722	0.000000000
saeRS	0.997017139	0.000000000
SaeS	0.999898610	0.000000000
sak	0.997017139	0.000000000
SarS	0.002366794	0.958714341
SarT	0.092117180	1.000000000
SarU	0.971181167	0.000000000
SarV	1.000000000	1.000000000
sdrC	0.000582424	1.000000000
spa	0.008757900	0.984250423
SplABCDEF	0.999567678	0.070103715
sspA	0.999601718	0.003676761
sspB	0.970896564	0.000165225
sspC	0.152550870	0.152590863

In this table the activity of the different nodes of the network in the two different steady states (steady state 1 [SS1] and steady state 2 [SS2]) is shown. The activation is indicated as values between 0 (inactive node) and 1 (node activation at maximum). Here only nodes are shown in which the activity, in one of the steady states, differs from zero. In the text of this chapter [chapter 3-1] you can find a detailed description of the two steady states.

By using the computer program SQUAD two different steady states were determined in this network. The first steady state (SS1) represents a more invasive, toxic phenotype in which for example different haemolysins (*hla; hlb; hld; hlgA; hlgB; hlgC*), proteases (*splA-F; aur*), a DNAse and *geh* a glycerol ester hydrolase is up-regulated. These nodes are known for producing toxins which are able to destroy erythrocytes, cleave proteins and DNA as well as esters of membrane-lipids and fat in adipose tissue. This makes these substances very tissue destructive and hence they are of great importance for invading new tissue. The second steady state detected (SS2) represents a biofilm producing phenotype in which for example *rot* (repressor of toxins), the clumping factor B (*clfB*) and the binding proteins *sdrC* and *spa* is up-regulated. These nodes are known to enhance the biofilm forming ability of the colony [13; 8]. (see figure: 3-1 and table: 3-3). A switch between these two steady states, the invasive (SS1) and the biofilm building (SS2), is accurately modelled *in silico*, simply by up-regulating or down-regulating the activity of the *agr*-locus, by increasing or decreasing the *AIP* level, simulating a surrounding with a high and with a low *SA* density respectively.

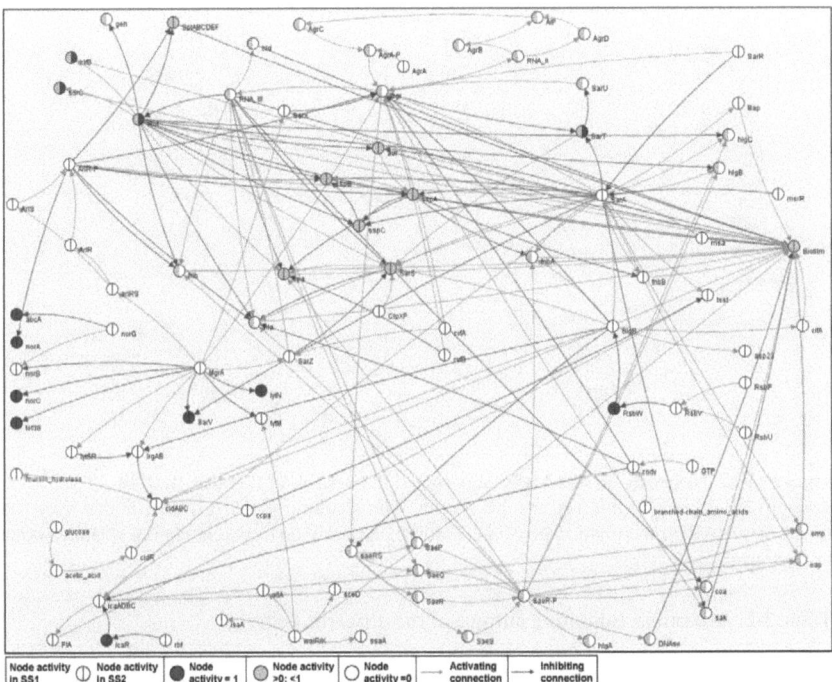

Figure 3-1: The two Steady states of the simulated network: Steady state 1 (SS1) represents a more invasive, noxious phenotype. Steady state 2 (SS2) represents a biofilm producing phenotype. The activation levels of the nodes in the corresponding two steady states are shown graphically. The left sides of the circles, illustrating the different nodes, are representing the activation levels of each node in steady state 1. The right sides of the circles are showing the activation level in steady state 2 (source: own data and picture; figure already shown in Audretsch et al. 2013).

3 – Experiments and results

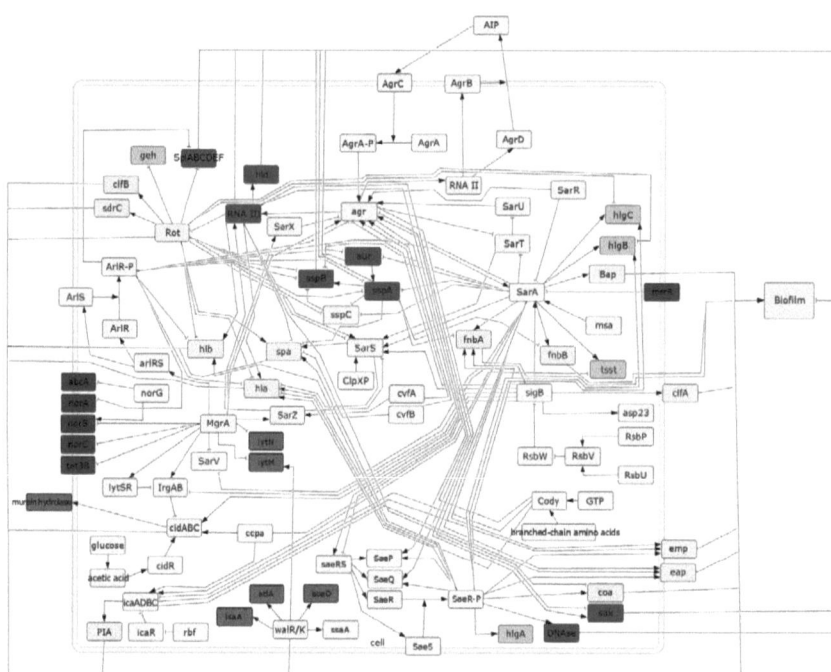

Figure 3-2: Network around the agr-locus: The different nodes and their activating respectively inhibiting interactions are shown. The shading of the different nodes show the effect they have on the phenotype of *SA*. (source: own data and picture; figure already shown in Audretsch et al. 2013).

Table 3-2: Activating/Inhibiting outputs of the different nodes:

Node	akt.output+ref	inh.output+ref	node	akt.output+ref	inh.output+ref
acetic acid	cidR[80]				
Agr	RNAIII[77]; RNAII[77]; SaeRS[77]; sak[52]	SarT[97]	Rot	biofilm[13]; sdrC[94]; clfB[94]; SarS[4]; spa[13]	geh[94]; SplA-F[94]; aur[44]; hlgB[13]; hlgC[13]; sspA[44]; sspB[44]; sspC[44]; hla[13]; hlb[13]
AgrA-P	agr[77]		RsbP	RsbV[80]	
AgrB	AIP[77]		RsbU	RsbV[80]	
AgrC	AgrA-P[77]		RsbV		RsbW[80]
AgrD	AIP[77]		RsbW		SigB[80]
AIP	AgrC[77]				
ArlR-P	SarA[34];	agr[34]; norA[33]; sspB[34]; sspA[34]; SplA-F[34]; hla[34]; hlb[34]			

3 – Experiments and results

arlRS	ArlS[34]; ArlR[34];	
ArlS	ArlR-P[34]	
Aur	sspA[75]	biofilm[8]
Bap	biofilm[107]	
branched-chain amino acids	cody[67]	
Ccpa	cidABC[102]; icaADBC[102]	
cidABC	murein hydrolase[119]; biofilm[88]	
cidR	cidABC[119]	
clfA	biofilm[5]	
clfB	biofilm[5]	
ClpXP	SarS[19]	
Coa	biofilm[79]	
Cody	icaADBC[67]; hla[67]; agr[67]	
cvfA	SarZ[73]; agr[73]	
cvfB	agr[73]	spa[73]
DNAse		biofilm[49]
Eap	biofilm[53]	
Emp	biofilm[53]	
fnbA	biofilm[8]	
fnbB	biofilm[8]	
Glucose	acetic acid[119]	
GTP	cody[67]	
hla	biofilm[14]	
hlb	biofilm[49]	
hld		biofilm[113]
icaADBC	PIA[8]; emp[53]; eap[53]	
icaR		icaADBC[23]
lrgAB		cidABC[119]
lytSR	lrgAB[13]	
MgrA	SarX[4]; agr[4]; SarZ[4]; lrgAB[50]; lytSR[50]; arlRS[50]	lytM[50]; lytN[50]; SarV[4]; norB[108]; norC[108]; tet38[108]
msa	SarA[95]	
msrR		SarA[91]
norG	norB[109]	abcA[109]
PIA	biofilm[8]	
rbf		icaR[23]

SaeR-P	SaeQ[1]; SaeP[1]; hla[66]; hlb[66]; spa[38]; fnbA[66]; hlgB[89]; hlgC[89]; emp[53]; eap[53]; coa[66]; hlgA[89]; DNAse[39]	
saeRS	saeP[1]; saeQ[1]; saeR[1]; saeS[1]	
SaeS	saeR-P[1]	
sak		biofilm[52]
SarA	agr[34]; hlgC[13]; hlgB[13]; tsst[13]; fnbB[13]; fnbA[13]; emp[53]; eap[53]; icaADBC[107]; hla[13]; SarS[19]; Bap[107]	sspC[44]; sspA[44]; sspB[44]; aur[44]; sarT[97]; sarV[4]; sak[52]
SarR	agr[4]	SarA[44]
SarS	spa[19]	hla[19]
SarT	SarS[98]	SarU[68]
SarU	agr[97]	
SarX		agr[4]
SarZ	RNAIII[4]	SarS[4]
sceD		
sdrC	biofilm[5]	
SigB	asp23[81]; clfA[32]; SarA[7]; SarS[13]; cidABC[87]; fnbA[32]	agr[7]; lrgAB[87]; saeRS[36];
spa	biofilm[8]	
SplA-F		biofilm[8]
sspA	sspB[75]	sspC[75]; fnbB[75]; fnbA[55]; biofilm[8]; spa[55]
sspB		biofilm[8]

3 – Experiments and results

RNA II	AgrB[77]; AgrD[77]; AgrA[77]		sspC		sspB[75]
RNA III	SaeQ[78]; SaeP[78]; hlb[13]; hld[13]; hla[34];	Rot[13]; coa[13]; spa[13]; SarS[19]; biofilm[8]	walR/K	isaA[28]; atlA[28]; lytM[28]; sceD[28]; ssaA[28]; biofilm[28]	

In this table all the nodes of the constructed network are shown. In the second and third column the corresponding activating and inhibiting outputs are listed. In brackets the corresponding references for the listed interactions are shown.

Creating a network, like the one presented here, holds many problems. One of these problems is for example selecting the cutout of a template network, as it can be found in nature with an infinite number of nodes. Thus selecting the nodes on the border and selecting which node to include and which node to exclude is not always easy. Yet when excluding a peripheral node it is important to keep the influence of this node on the whole system in mind and include this influence, as external factor, in the simulations, done later on with this network. Finding system states in a network, like the ones described above, is also not that easy as it sounds. When for example looking at a network like the one presented here with 94 nodes, there are 2^{94} possible combinations when just comparing the states 0 and 1 for each node. The more precise the simulation should be the more increments between 0 and 1 are needed and thus the amount of possible combinations tends to become infinite. From this large amount of system states the interesting ones need to be selected. The interesting states are generally the stable ones, to which the system comes back after temporally limited external inputs and in which the expression of each node is constant. It is believed and seems obvious that only states that are present for a certain amount of time might have a noteworthy influence. Yet there is also a dynamic environment, providing different, some times also constant, external inputs. Such constant external inputs lead to a change in the whole network, resulting in a change of the activation of the different nodes and thus to different steady states. Thus a lot of knowledge about the external factors, as well as about the steady states and their link to the phenotypic output and also a lot of experimental work *in silico* is inalienable to find system states, fitting the variables, found in the *in vitro* experiments.

3.2 - Comparative microarray analysis (consistence of simulations with previous knowledge)

3 – Experiments and results

To figure out the consistence of previously published data with the simulations and to show that the network not only includes what is currently known about the nodes and their interaction, yet also reacts as one would expect it from earlier publications, a microarray data analysis was conducted. The consistency was investigated between *in silico* and *in vitro* in an agr^+ versus agr^- as well as in a $sarA^+$ versus $sarA^-$ and in a biofilm versus planctonic scenario. First of all *in silico* agr^+ versus agr^- in the exponential and the post exponential growth phase was compared to the *in vitro* agr^+ versus agr^- of an *UAMS-1 SA* strain, once in the exponential phase (OD 1.0 at 560nm) and once in the post exponential phase (OD 3.0 at 560nm) [17]. In this analysis a consistency between *in silico* and *in vitro* results of 81.43% and 71.43% respectively was acquired. Yet *in silico* agr^+ versus agr^- was also compared to agr^+ versus agr^- in a *RN27 SA* strain in the stationary growth phase [30]. Here an *in vitro in silico* consistency of 80.00% was acquired.

Secondly *in silico* $sarA^+$ versus $sarA^-$ again in the exponential and the post exponential growth phase was compared to *in vitro* $sarA^+$ versus $sarA^-$ of an *UAMS-1 SA* strain in the exponential phase (OD 1.0 at 560nm) and in the post exponential phase (OD 3.0 at 560nm) [17]. Here an *in vitro in silico* consistencies of 71.43% and 75.71% respectively was acquired. In this scenario *in silico* $sarA^+$ versus $sarA^-$ was also compared to $sarA^+$ versus $sarA^-$ in a *RN27* strain in the stationary growth phase [30]. Here an *in vitro in silico* consistency of 68.57% was acquired.

In the third scenario two *in silico* situations were compared to three *in vitro* situations. The first *in silico* situation was not biofilm forming SS2 versus biofilm forming SS1 which first was compared to a late exponential phase (6h) planctonic culture versus a maturing (48h) biofilm culture [12] (*in vitro in silico* consistency of 57.75%). Secondly it was compared to an, in biofilm grown (for 24hrs) *SA113* versus a planctonically grown (for 24hrs) *SA113* colony [86]. Here an *in vitro in silico* consistency of 56.34% was acquired. Third this first *in silico* situation (SS1 vs. SS2) was compared to an *in vitro* situation with an *UAMS-1 SA* colony, grown in biofilm (one week old, grown in flow cell) versus stationary phase (OD 3.5 at 560nm), planctonically grown *UAMS-1 SA* colony [6] (*in vitro in silico* consistency of 57.75 %). The second *in silico* situation was a biofilm forming phenotype with low *AIP* concentrations versus a biofilm negative phenotype, induced by high *AIP* concentrations. This second *in silico* situation was also compared to all three *in vitro* situations with *in silico in vitro* consistencies of 76.06%, 71.83% and 77.46% respectively. (see also table 3-3 with selected nodes or the full table (table 7-2) with all compared nodes in the supplementary material).

3 – Experiments and results

In all scenarios there are nodes that didn't show the same reaction as in the simulation. Yet there is not one node that reacts inconsistent in all scenarios. Concerning the agr^- scenarios, the only nodes that react inconsistent in all three datasets are *AgrB, fnbA, hlb, saeR, sak* and *sarU*. *Aur, fnbB, isaA-D, sspA-C* and *tsst* are the only inconsistent nodes in all three *sarA*- datasets. The only nodes that reacted inconsistent in all biofilm$^+$ vs. biofilm$^-$ datasets are *agrB, agrD, aur, saeR, sak, splABDEF*. Moreover one can see on the basis of the two different *in silico* biofilm$^+$ vs. biofilm$^-$ datasets and the different *in vitro* $sarA^-$, $agrA^-$ and biofilm datasets, that already small changes in the scenario can cause the inconsistency of some nodes. Thus, with more insight in the the real parameters of the scenarios (e.g. growth conditions, nutrient availability, and pH) an even higher consistency with the simulation needs to be proposed, given that the simulation has the capability to incorporate the variables.

3 – Experiments and results

Table 3-3: (A, B, C) Comparative microarray analysis:

A)

node	agrA⁺ vs. agrA⁻							
	A1: Cassat agrA OD 1	A1-T1 correlation	agrA up/down sim T1	A2: Cassat agrA OD 3	A2-T3 correlation	agrA up/down sim T3	A3: Dunman agrA RN27	A3-T3 correlation	agrA up/down sim T3
AgrA	x		x	x		x	x		x
arlR	=		=	=		=	=		=
arlS	=		=	=		=	=		=
Rot	=		=	=		+	=		=
SaeR	=		+	=		+	=		+
SaeS	=		=	=		=	=		=
SarA	=		=	=		=	=		=
SigB	=		=	=		=	=		=
Number of compared nodes		70			70			70	
concordant nodes		57			50			56	
non concordant nodes		13			20			14	
in vitro in silico consistency in %		81.43			71.43			80.00	

3 – Experiments and results

B)

	sarA⁺ vs. sarA⁻							
node	B1: Cassat sarA OD 1	B1-T1 correlation	sarA up/down sim T1	B2: Cassat sarA OD 3	B2-T3 correlation	sarA up/down sim T3	B3: Dunman agrA RN27	sarA up/down sim T3
AgrA	=		=	=		=	+	=
arlR	=		=	=		=	=	=
arlS	=		=	=		=	=	=
Rot	=		=	=		=	=	=
SaeR	=		+	=		=	=	=
SaeS	=		=	=		=	=	=
SarA	−		×	×		×	×	×
SigB	=		=	=		=	=	=
Number of compared nodes		70			70			70
concordant nodes		50			53			48
non concordant nodes		20			17			22
In vitro in silico consistency in %		71.43			75.71			68.57

3 – Experiments and results

	Biofilm vs. Planctonic										
node	Biofilm Vs Planctonic (SS) Sim. 1	Biofilm Vs Planctonic (AIP) Sim. 2 (compared to Sim 1)	C1-Sim 1 correlation	C1 Sim 2 correlation	C1 Biofilm Vs Planctonic (maturing)	C2-Sim 1 correlation	C2-Sim 2 correlation	C2 Biofilm Vs Planctonic 24hr	C3-Sim 1 correlation	C3-Sim 2 correlation	C3 Biofilm Vs Planctonic OD 3.5
AgrA	=	=			=			=			=
arlR	=	=			=			=			=
arlS	=	=			=			=			=
Rot	+	−			=			=			=
SaeR	−	−			=			=			
SaeS	−				=			=			=
SarA	=	=			=			=			=
SigB	=	=			=			=			=
Number of compared nodes			71	71		71	71		71	71	
concordant nodes			41	54		40	51		41	55	
non concordant nodes			30	17		31	20		30	16	
in vitro in silico consistency in %			57.75	76.06		56.34	71.83		57.75	77.46	

In (a) three *in vitro* $AgrA^+$ vs. $AgrA^-$ scenarios are compared to an *in silico* $AgrA^+$ vs. $AgrA^-$ scenario. Then in (b) three $SarA^+$ vs. $SarA^-$ scenarios are compared to an *in silico* $SarA^+$ vs. $SarA^-$ scenario. Furthermore in (c) three *in vitro* biofilm forming vs. not biofilm forming scenarios are compared to two *in silico* biofilm forming vs. not biofilm forming scenarios. A black shading in one of the correlation columns shows that this node was not included in the analysis, because this node is the one that was changed externally to get the different scenarios. The intermediate gray shading in one of the correlation columns means that this node did not show the same reaction in the *in vitro* and the *in silico*. The dark gray shading in one of the correlation columns means that this node showed no difference between *in vitro* and *in silico*. All other shadings are just for better visualisation of the compared groups. In this table a "+" means that this node is up-regulated by the three fold in the wild type strain or by the 2.5 fold in the biofilm forming situation. A "−" means that this node is up-regulated by the three fold in the mutant strain or by the 2.5 fold in the not biofilm forming situation. Here only a few selected nodes are shown, for the full table see table 7-2 in the supplementary material. A detailed description of the scenarios can also be found in the Materials and Methods [chapter 2.2]; a complete table, with all nodes of the network can be found in the Apendix section, table 7.2.

3 – Experiments and results

Table 3-4: Microarray analysis summary:

	In vitro results	In-silico results	No. of nodes compared	No. of Concordant nodes	No. of non-concordant nodes	Consistency between in-vitro and in-silico results (%)
agrA⁺ vs. agrA⁻	agr⁺ vs. agr⁻ OD1	agrA up/down Simulation T1	70	57 eg.: arlR; Rot; SaeS; SarA; SigB.	13	81.43
	agr⁺ vs. agr⁻ OD3	agrA up/down simulation T3	70	50 eg.: arlR; Rot; SaeS; SarA; SigB.	20	71.43
	agr⁺ vs. agr⁻ RN27	agrA up/down simulation T3	70	56 eg.: arlR; Rot; SaeS; SarA; SigB.	14	80.00
sarA⁺ vs.sarA⁻	sarA⁺ vs. SarA⁻ OD1	sarA up/down simulation T1	70	50 eg.: AgrA; arlR; Rot; SaeS; SigB.	20	71.43
	sarA⁺ vs. SarA⁻ OD3	sarA up/down simulation T3	70	53 eg.: AgrA; arlR; Rot; SaeS; SigB.	17	75.71
	sarA⁺ vs. SarA⁻ RN27	sarA up/down simulation T3	70	48 eg.: arlR; Rot; SaeR; SaeS; SigB..	22	68.57

	In vitro results	In-silico results	No. of nodes compared		No. of Concordant nodes		No. of non-concordant nodes		Consistency between in-vitro and in-silico results (%)	
			Sim 1 SS2 vs. SS1	Sim 2 AIP low vs. AIP high	Sim 1 SS2 vs. SS1	Sim 2 AIP low vs. AIP high	Sim 1 SS2 vs. SS1	Sim 2 AIP low vs. AIP high	Sim 1 SS2 vs. SS1	Sim 2 AIP low vs. AIP high
Biofilm vs. Planctonic	Biofilm Vs Planktonic (maturing)	Biofilm Vs Planktonic (SS) Sim. 1	Biofilm Vs Planktonic (AIP) Sim. 2	71	71	41 eg.: arlR; SaeS; SigB	54 eg.: Rot; SaeS; SigB	30	17	57.75	76.05
	Biofilm Vs Planktonic 24hr	Biofilm Vs Planktonic (SS) Sim. 1	Biofilm Vs Planktonic (AIP) Sim. 2	71	71	40 eg.: AgrA; arlR; SaeS; SigB	51 eg.: AgrA; Rot; SaeS; SigB	31	20	56.34	71.83
	Biofilm Vs Planktonic OD 3.5	Biofilm Vs Planktonic (SS) Sim. 1	Biofilm Vs Planktonic (AIP) Sim. 2	71	71	41 eg.: AgrA; arlR; SaeS; SigB	55 eg.: AgrA; Rot; SaeS; SigB	30	16	57.75	77.46

Here an overview of the results from the mircroarray analysis is shown (table already shown in Audretsch et al. 2013).

3 – Experiments and results

Besides the tables, showing the *in silico in vitro* correlation of wild type vs. *agrA⁻*, wild type vs. *sarA⁻* and biofilm⁺ vs. biofilm⁻ graphics are shown where one can see, by means of little flags that the expression of most of the nodes under *saeRS⁻* and *agr⁻* (see figure: 3-5 and figure: 3-3) conditions is the same as one would expect it from the knowledge we have from different previous publications). To get an even deeper insight in the reaction of different nodes also time courses were created, where at 50sec either *saeRS* (see figure: 3-6) or *agr* (see figure: 3-4) were knocked-out. Here, by means of little flags, also the consistency of the reaction from different nodes of the simulation, with the knowledge from different previous publications is shown.

3 – Experiments and results

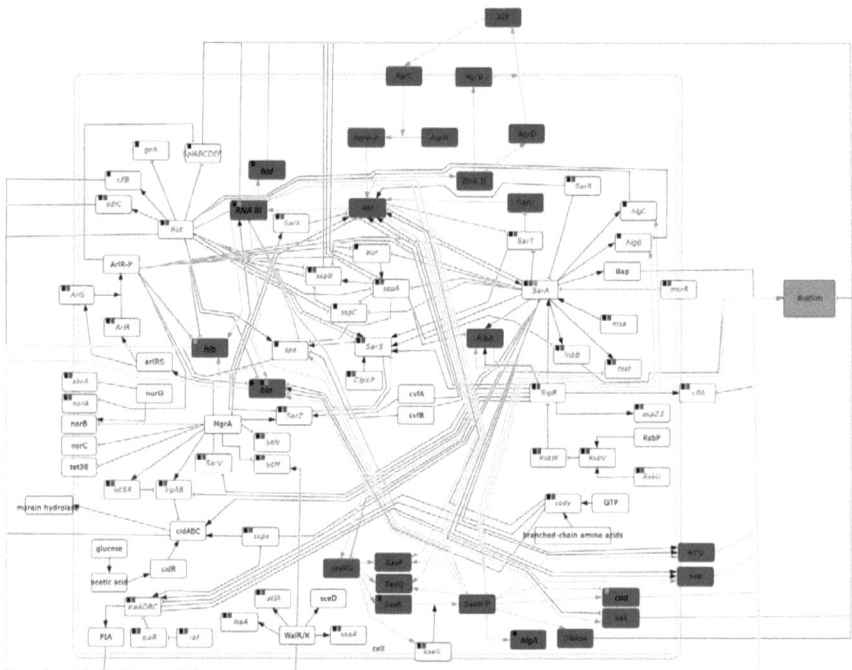

Figure 3-3: Impact of *agr⁻* on the network: In this figure the qualitative difference in the activation level of all the nodes of the network, between a simulated *wild type* and a simulated *agr⁻* mutant strain are shown. A difference between *wild type* and mutant strain was adopted when the activation strength in the mutant strain was at least 2.5 fold higher or lower than in the *wild type* strain. Shaded in dark gray, all the nodes are shown that are down-regulated in the mutant strain, when comparing it to the *wild type* strain. Shaded in light gray on the other hand, the up-regulated nodes are shown. Coloured in white, nodes are shown that remain unaffected. Furthermore in the figure, with little flags the reference is indicated in which, under the same circumstances, an equal reaction of the corresponding node can be found. In addition the names of all the nodes, whose reaction is verified by reference, are written in bold italics (source: own data and picture; figure already shown in Audretsch et al. 2013).

3 – Experiments and results

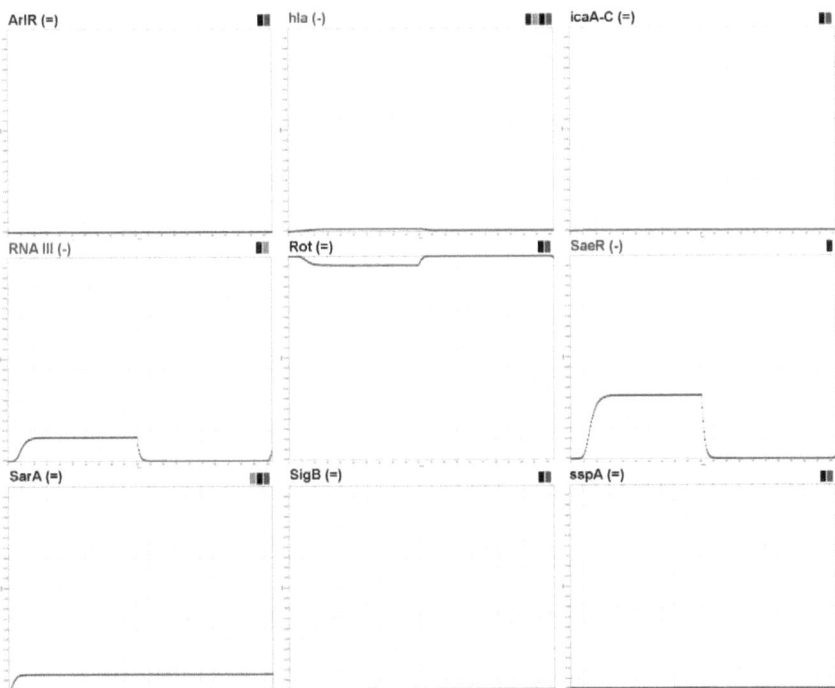

Figure 3-4: Reaction of different nodes, when knocking-out *agr in silico*: In this figure the simulated reaction of *ArlR, hla, icaA-C, RNA III, Rot, SaeR, SarA, SigB* and *sspA* as a reaction to an *agr* knocked-out pertubation at 50 sec are shown. The abscissae indicate the time in seconds and the ordinates indicate the activation level of the node. When, by knocking out *agr*, the activation level was reduced to a level more then 2.5 times lower than in the *wild type* and thus the node was adopted to be down-regulated *in silico*, the names of the nodes are written in light gray. Written in black are the names of all the unaffected nodes. Furthermore in the figure with differently shadeded, little flags the reference is indicated in which, under the same circumstances, an equal reaction of the corresponding node can be found (source: own data and picture; figure already shown in Audretsch et al. 2013).

3 – Experiments and results

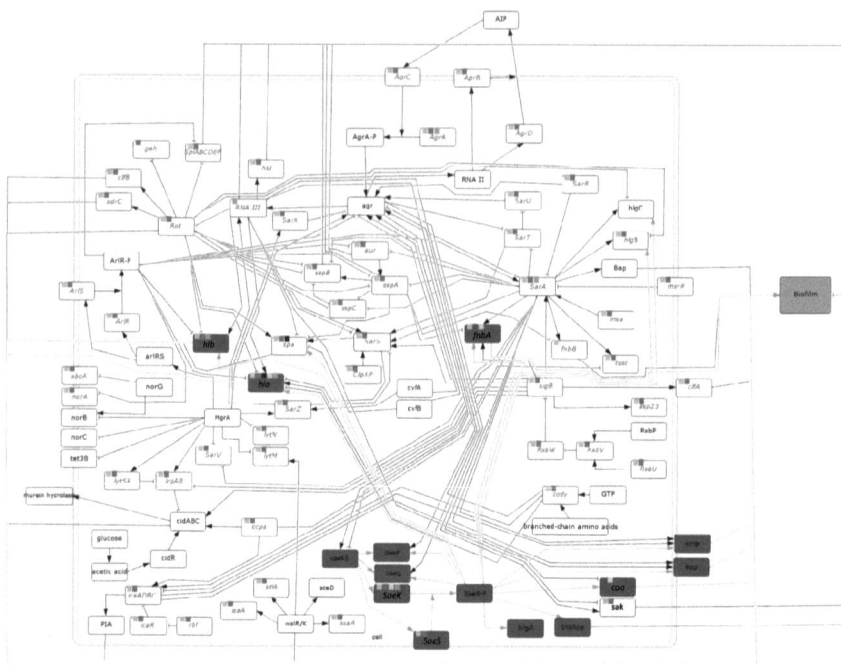

Figure 3-5: Impact of *saeRS⁻* on the network: In this figure the qualitative difference in the activation level of all the nodes of the network, between a simulated *wild type* and a simulated *saeRS⁻* mutant strain are shown. A difference between *wild type* and mutant strain was adopted when the activation strength in the mutant strain was at least 2.5 fold higher or lower than in the *wild type* strain. Shadeded in dark gray, all the nodes are shown that are down-regulated in the mutant strain, when comparing it to the *wild type* strain. Shadeded in light gray on the other hand, the up-regulated nodes are shown. Coloured in white, nodes are shown that remain unaffected. Furthermore in the figure, with little flags the reference is indicated in which, under the same circumstances, an equal reaction of the corresponding node can be found. In addition the names of all the nodes, whose reaction is verified by reference, are written in bold italics (source: own data and picture; figure already shown in Audretsch et al. 2013).

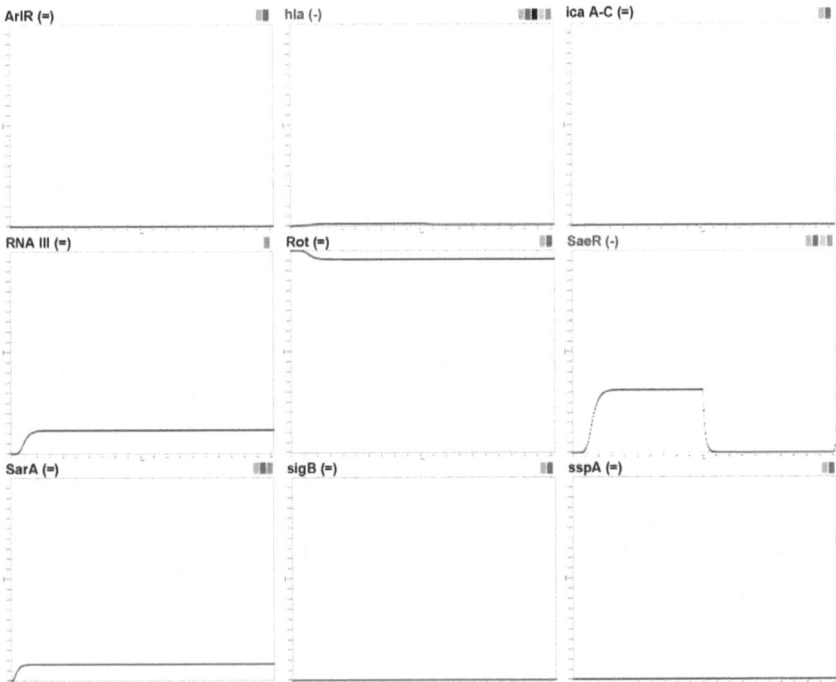

Figure 3-6: Reaction of different nodes when knocking-out *saeRS in silico*: In this figure the simulated reaction of *ArlR, hla, icaA-C, RNA III, Rot, SaeR, SarA, SigB* and *sspA* as a reaction to a *SaeRS* knocked-out pertubation at 50 sec is shown. The abscissae indicate the time in seconds and the ordinates indicate the activation level of the node. When, by knocking out *SaeRS*, the activation level was reduced to a level more then 2.5 times lower than in the *wild type* and thus the node was adopted to be down-regulated *in silico*, the names of the nodes are written in light gray. Written in black are the names of all the unaffected nodes. Furthermore in the figure with differentially shaded, little flags the reference is indicated in which, under the same circumstances, an equal reaction of the corresponding node can be found (source: own data and picture; figure already shown in Audretsch et al. 2013).

3.3 - Testing the mutants
(in vitro consistence with in silico predictions)

3.3.1 - Northern blots

To test for the robustness of the network it was tested, whether it could be used for predictions about alterations in the signalling cascades and for predictions about changes in the phenotypic reactions of *SA* colonies, dependent on changes of different nodes in the network. Therefore five different knock-out mutants (*agr⁻, sae⁻, sae⁻/agr⁻, sigB⁺, sigB⁺/sae⁻*) were used and the expression of different prominent nodes (*asp (sigB), sae, sarA, agr*) was

3 – Experiments and results

investigated by conducting a Northern blot (figure 3-8). These nodes were selected for the knock-out mutants, because of their proposed importance for the *QS* [13; 77; 89] and because of their importance for the whole network that was set up. The importance, the different nodes have for the network, was estimated simply by supposing that nodes with more connections must have a greater impact on the reaction of the whole network and thus need to be of bigger importance for it. Thus, just the amount of interactions the nodes have with other nodes in the network was compared, for selecting putatively more important nodes from the more unimportant ones.

The *agr*-locus has 24 interacions with other nodes of the network. As a result, together with *sarA*, *agr* is the node with the most interactions of the network. Through these connections the *agr*-locus up-regulates for example itself in a positive feedback loop [77]. Also *sae* [77] and different haemolysins (*hla; hlb; hld*) are up-regulated by *agr* [13; 34]. The *agr*-locus on the other hand down-regulates for example *rot* [13]. When the *agr*-locus is knocked-out, this results in an enhanced biofilm forming ability [8].

SarA has, like the *agr*-locus, 24 connections to other nodes in the network, what also reflects the importance of this node. Nodes that are up-regulated by *SarA* are for example the *agr*-locus [34], different haemolysins (*hla; hlgB; hlgC*) [13], the toxic shock syndrome toxin (*TSST*) [13] and the intracellular adhesion proteins A-C (*icaA; icaB; icaC*) [107]. Down-regulated on the other hand are for example the serine proteases *sspA, sspB, sspC* [44].

Rot, the abbreviation for repressor of toxins is a node with 16 connections to other nodes in this network and thus it has rank two, when listing all the nodes by quantity of connections with other nodes. The serine proteases *sspA, sspB* and *sspC* [44] and many haemolysins such as *hlgb, hlgc, hla, hlb* are down-regulated by *rot* [13]. The biofilm forming ability on the other hand is up-regulated by *rot* [13].

The *sae*-locus has many up-regulating connections. For example many haemolysins (*hla; hlb; hlgA; hlgB; hlgC*) [66; 89] and the fibronectin binding protein A (*fnbA*) are up-regulated [66]. All in all the *sae*-locus has 12 interactions and has thus rank three with the third most connections in the network.

Of great importance for the *SA* stress response is the alternative sigma factor *SigB* of the *SA* RNA polymerase [80]. Here *sigB* has 10 interactions with other nodes of the network. When listing all the nodes by quantity of connections *sigB* has the fourth most connections. *SarA* is for example up-regulated by *SigB* [7]. *SigB* also up-regulates the murein hydrolase activators *cidA, cidB* and *cidC* [87], which are also known to contribute in biofilm formation [88].

The nine interactions with other nodes reserves rank 5 for the *arl*-locus, when estimating the impact a node has on the network, by using the quantity of thir connections. Like many other nodes of the network the *arl*-locus contributes for example in up-regulating *sarA* [34]. Down-regulated by the *arl*-locus are for example again many haemolysins (*hla; hlb*) as well as the serine proteases *splA-F* [34] and the *agr*-locus [34].

The expression data obtained from the Northern blots were compared to the corresponding simulated knock-out mutants, created by down-regulating the knocked-out nodes in SQUAD, leading as in the *in vitro* knock-out mutants to a near to zero expression of the node. Moreover the different TCS incorporated in the simulation were up-regulated a little bit (see the large bar in figure: 3-7) to simulate an *in vitro* like surrounding, in which all the TCS are assumed to be stimulated to some extend. The activation level of these prominent nodes was in the *in vitro* mutants and in the simulated mutants qualitatively comparable.

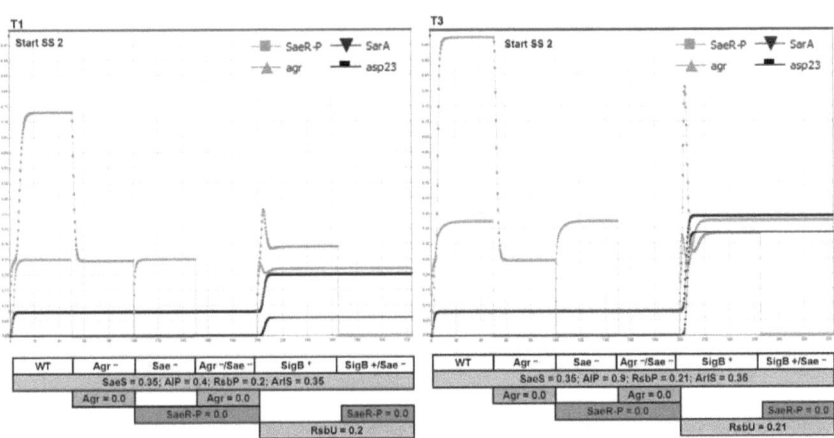

Figure 3-7: Northern blot simulations: In this figure the in-silico expression of *sarA, agr, saeR-P*, the phosphorylated and thus activated *saeR* and *asp23*, the surrogate for the *sigB* expression, in the 6 tested strains (*WT* [0-50], *agr⁻* [50-100], *sae⁻* [100-150], *sae⁻/agr⁻* [150-200], *sigB⁺* [200-265], *sigB⁺/sae⁻* [265-320]) is shown at the T1 and T3 growth phase. Moreover the SQUAD-pulses, programmed for the SQUAD-simulation, are shown (source: own data and picture; figure already shown in Audretsch et al. 2013).

3 - Experiments and results

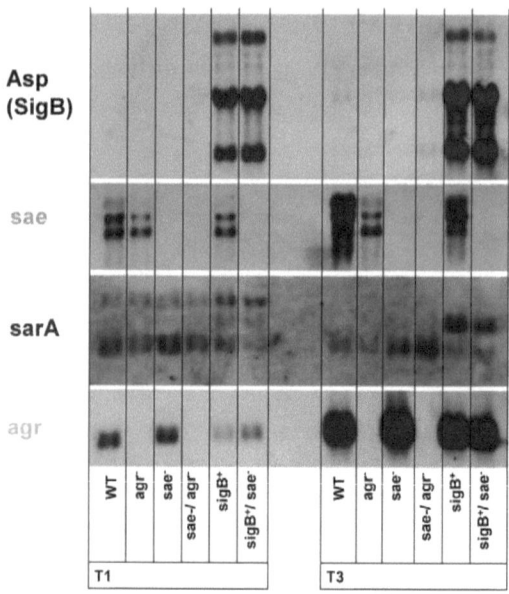

Figure 3-8: Northern blot: This figure shows the Northern blot, made with the *WT* strain and the five different strains with mutations in important nodes. In all strains the expression of *agr, sarA, sae, and asp (sigB)* is detected in T1 and T3 (source: own data and picture; figure already shown in Audretsch et al. 2013).

3.3.2 - Biofilm strength

To test whether the biofilm forming capability of these mutants is in the same way affected as the biofilm forming capability of the simulated mutants, a biofilm adherence assay was made (see figure: 3-9). All the simulated mutants in which the biofilm intensity was higher than 0.5 were noted as biofilm forming phenotypes. Just as well as the RNA expression, obtained from the Northern blots, the biofilm forming ability, obtained from the biofilm adherence assays, was in the *in vitro* and in the *in silico* scenario qualitatively affected in the same way. All the mutant strains were able to build a biofilm; only in the *wild type* strain the biofilm forming capability was impaired. To my knowledge not shown until now and very unexpected was the extensive biofilm forming ability of the *sae⁻* strain, which were found in the *in silico* simulation and also in the *in vitro* biofilm adherence experiments. To make sure that this result is not a side effect of the *sae⁻* mutation construction also biofilm adherence assays with the *sae⁻* complementation strain were conducted, in which the biofilm forming ability was found impaired again. This suggesting that *sae* is an important factor in the biofilm regulation mechanism. More precisely: *sae* plays a crucial role in detering *SA* from building biofilms in inappropriate situations and maybe is even crucial, besides *agr*, for the dissemination of *SA* from an already formed biofilm.

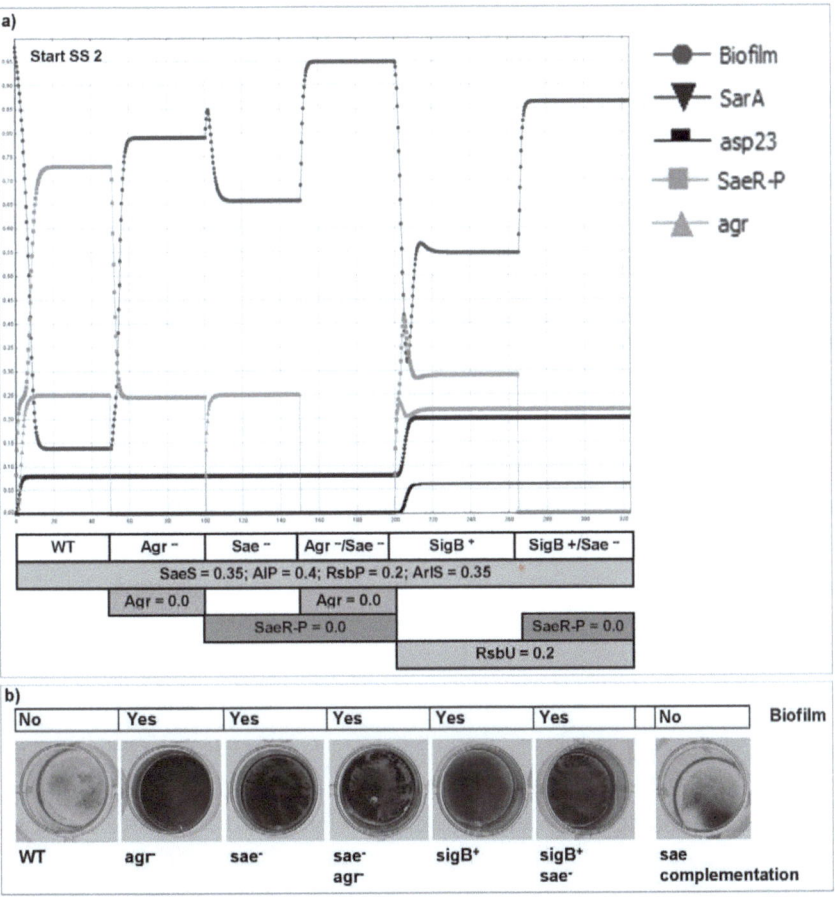

Figure 3-9: Biofilm adherence assay: a) In this figure the in-silico expression of *sarA*, *agr*, *saeR-P*, the phosphorylated and thus activated *saeR* and *asp23*, the surrogate for the sigB activation level and also the biofilm formation strength in the 6 tested strains (*WT* [0-50], *agr⁻* [50-100], *sae⁻* [100-150], *sae⁻/agr⁻* [150-200], *sigB⁺* [200-265], *sigB⁺/sae⁻* [265-320]) is displayed. Moreover the SQUAD-pulses, programmed for the SQUAD-simulation, are shown. **b)** Here the in-vitro biofilm results of the different strains are shown. The biofilm forming ability is qualitatively comparable to the in-silico experiments. Moreover the biofilm forming ability of the *sae⁻* complementation is shown. The biofilm intensity is comparable to the *WT* hence the biofilm up-regulation in *sae⁻* can de facto be ascribed to the *sae* mutation (source: own data and picture; figure already shown in Audretsch et al. 2013).

3 – Experiments and results

3.3.3 - Importance of *Sae* for biofilms

3.3.3.1 - *Venn-diagrams*

Until now it was not shown, that knocking-out *sae* in *SA* leads to enhanced biofilm formation, compared to the *wild-type SA* strain and that thus *sae* needs to be proposed to inhibit the biofilm forming ability of *SA*. To get a hint which of the many genes of *SA* could be responsible for this effect it was just proposed, that the responsible gene should be one of those with a different expression strength first when comparing biofilm forming and planktonic living strains and second when comparing sae^+ and sae^- strains. The gene of interest should first be up-regulated by *sae* and thus can lead to weaker biofilms in case of an activated *sae*-locus. Second this gene should be down-regulated in biofilms given it usually is responsible for impairing the biofilm building ability of *SA*.

In the Venn-diagrams 14 genes were found, changed in *wt* vs. sae^- and planktonic vs. biofilm scenarios (see table 3-5). The gene *SA1007* an *Alpha-Haemolysin precursor* was found in 12 of the 50 Venn-diagrams, yet there was no consistency whether this gen is up- or down-regulated in biofilms. The gene found, with nine times, the second most was *SA1755* a hypothetical protein, which is up-regulated by *sae* and down-regulated in biofilms. *SA2206* the *IgG-binding protein SBI* was found in four Venn-diagrams and is up-regulated by *sae*, yet it is also up-regulated in biofilms which means, that it is unlikely that this gene is responsible for the biofilm inhibiting Effect of *sae*. Also up-regulated by *sae* and in biofilms is the, two times in the Venn-diagrams found, gene *SA0219,* coding for a *formate acetyltransferase activating enzyme*. Also two times, yet fulfilling the search criteria (up-regulated by *sae* and down-regulated in biofilms) the two hypothetical genes *SA0224* (ORFID: *SA0224~ hypothetical protein*, similar to *3-hydroxyacyl-CoA dehydrogenase*) and *SA0357* (ORFID: *SA0357~ hypothetical protein*, similar to *exotoxin 2*) were found. Also fulfilling the search criteria, yet not hypothetical and also found two times in the Venn-diagrams are the genes *SA0744* (*extracellular ECM* and *plasma binding protein*) and *SA0746* (*staphylococcal nuclease*). Moreover there were four hypothetical genes, not fulfilling the search criteria, each of which was found one time. These genes are *SA0213* (*conserved hypothetical protein*), *SA0663* (*hypothetical protein*), *SA1000* (ORFID: *SA1000~ hypothetical protein*, similar to *fibrinogen-binding protein*) and *SA1709* (ORFID: *SA1709~ hypothetical protein,* similar to *ferritin*). Furthermore there were two genes; also found one time in the Venn-diagrams, yet these genes fulfilled the search criteria. These genes are *SA1271* (ORFID: *SA1271~threonine deaminase IlvA homolog*) and the hypothetical protein *SA0394*.

3 – Experiments and results

In summary the most found, the search criteria fulfilling, not hypothetical and thus most promising genes for further analysis were *SA0744* (*extracellular ECM and plasma binding protein*) and *SA0746* (*staphylococcal nuclease*). Thus both of them are supposed to be of importance for the link between the *sae*-locus and the biofilm formation ability.

3 – Experiments and results

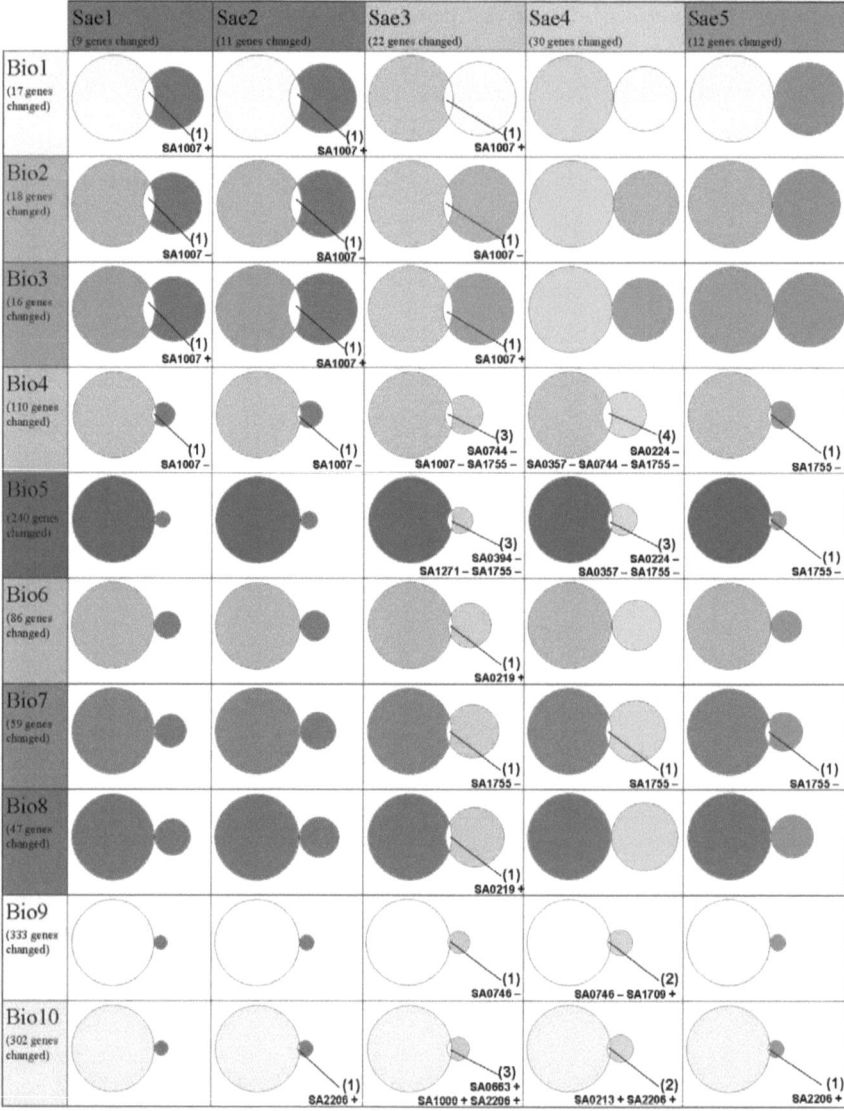

Figure 3-10: Venn-diagrams comparing genes, differentially expressed in *wt* vs. *sae⁻* and under planctonic vs. biofilm conditions: In these Venn-diagrams ten microarray experiments, evaluating the difference in gene expression of *SA* planctonically grown vs. grown in biofilm (Bio1-Bio10) are compare to five microarray experiments evaluating the difference in gene expression in *wt* vs. *sae⁻* (Sae1-Sae5). Genes differentially expressed in both scenarios are always up-regulated in the *sae* experiments. The plus and minus show whether this gene is up- or down-regulated in the biofilm (source: own picture).

Table 3-5: Genes changed in *wt* vs. *sae⁻* and planctonic vs. biofilm scenarios:

Gene	Amount	Gene Description	UP/DOWN
SA1007	12	Alpha-Haemolysin precursor	sae+/bio1 & 3+; bio2 & 4–
SA1755	9	hypothetical protein (CHIPS)	*sae+/bio–*
SA2206	4	IgG-binding protein SBI	sae+/bio+
SA0219	2	formate acetyltransferase activating enzyme	sae+/bio+
SA0224	2	ORFID: SA0224~ hypothetical protein, similar to 3-hydroxyacyl-CoA dehydrogenase	*sae+/bio–*
SA0357	2	ORFID: SA0357~ hypothetical protein, similar to exotoxin 2	*sae+/bio–*
SA0744	2	extracellular ECM and plasma binding protein	*sae+/bio–*
SA0746	2	staphylococcal nuclease	*sae+/bio–*
SA0213	1	conserved hypothetical protein	sae+/bio+
SA0394	1	hypothetical protein	*sae+/bio–*
SA0663	1	hypothetical protein	sae+/bio+
SA1000	1	ORFID: SA1000~ hypothetical protein, similar to fibrinogen-binding protein	sae+/bio+
SA1271	1	ORFID:SA1271~threonine deaminase IlvA homolog	*sae+/bio–*
SA1709	1	ORFID: SA1709~ hypothetical protein, similar to ferritin	sae+/bio+
Sum 14	Sum 41		

This table shows the genes changed in *wt* vs. *sae⁻* and planctonic vs. biofilm scenarios. The "Amount" column shows in how many Venn-diagrams the gene can be found changed in both scenarios. Also a gene description can be seen, as well as the type of expression change, either up- or down-regulated in the respective scenario. Genes fulfilling the search criteria (up-regulated by *sae* and down-regulated in biofilms) are written in bold and italic.

3.3.3.2 - *DNAse concentration in biofilms*

It is known that free nucleic acids play an important role in biofilm formation [63]. Moreover *DNAse* itself helps dissolving biofilms [49] and is up-regulated by *sae* [39]. Furthermore staphylococcal nuclease is one of the genes found in the Venn-diagrams as up-regulated by *sae* and down-regulated in biofilms. Also in the *in silico* experiments, conducted for this thesis, it was shown that when knocking-out *DNAse* a *sae⁺* strain is producing a biofilm. On the other hand a *sae⁻* strain is not producing biofilms when *DNAse* is complemented (see figure 3-11). Thus the *in silico* experiments too suggest *DNAse* to be important for the biofilm building impairment in *sae⁺ SA* strains. To see if this hypothesis bears the comparison with the real world first of all *in vitro* tests for qualitatively evaluating the amount of produced *DNAse* in different *SA* strains were done.

Each strain was measured nine times. In these experiments a clear (significant) difference between the four strains was found. The *wild-type* strain showed the strongest *DNAse* production with an average areola diameter size of 1.5cm (comparable to 300U/ml), followed by the *agr⁻* strain with an average areola diameter size of 1.38cm (comparable to 100U/ml). The next strongest *DNAse* producer with an average areola diameter size of 0.6cm was the

3 – Experiments and results

sae⁻ strain. The weakest *DNAse* production with an average areola diameter size of 0.14cm showed the *agr⁻/sae⁻* double mutant strain.

This means, that the *agr*-locus and the *sae*-locus each play an important role in activating the *DNAse* production. Yet the *sae*-locus showed a much stronger impact. When knocking-out both nodes the *DNAse* production was nearly down to zero. When simulating the *DNAse* expression under *sae⁻*, *agr⁻* and *agr⁻/sae⁻ in silico* qualitatively the same results were acquired (see figure 3-13). All this suggests that, like in the model the *sae*-locus has a very direct effect on the *DNAse* production. *Agr* on the other hand has a weaker maybe indirect, maybe on *sae* dependant effect on the *DNAse* production.

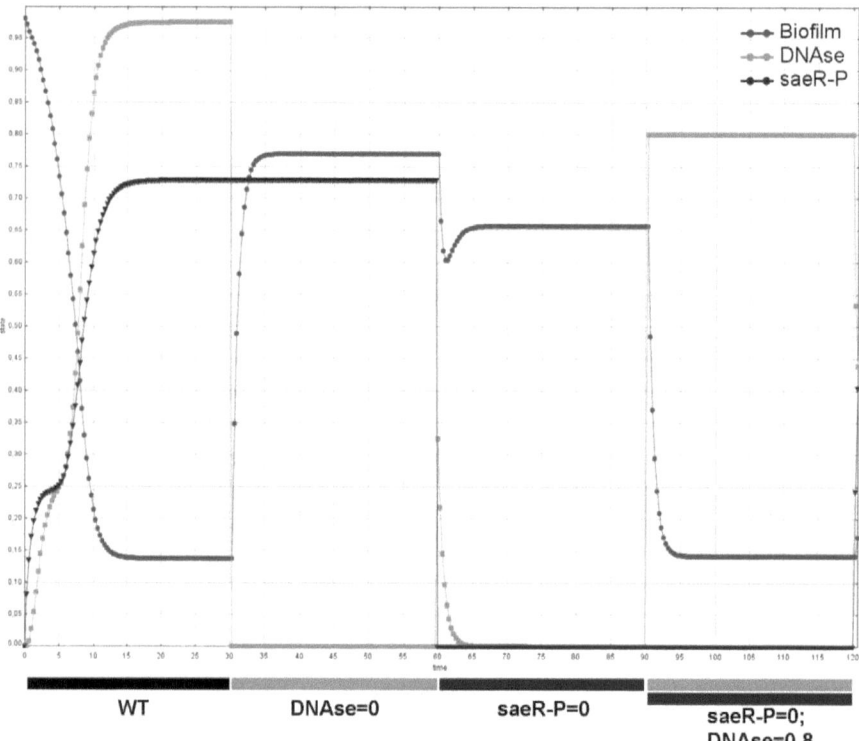

Figure 3-11: *DNAse* effect on the biofilm building ability of *wt* and *sae⁻* strains *in silico*: The effect of *DNAse* on the biofilm building ability of *SA* was tested *in silico* by comparing a *wt SA* strain to a *DNAse* knock-out strain, a *sae⁻* strain and a *sae⁻* strain with *DNAse* complementation. It is obvious that the biofilm building ability *in silico* depends mainly on the *sae* regulated *DNAse* level and not that much on the *sae* activity itself (source: own data and picture).

3 – Experiments and results

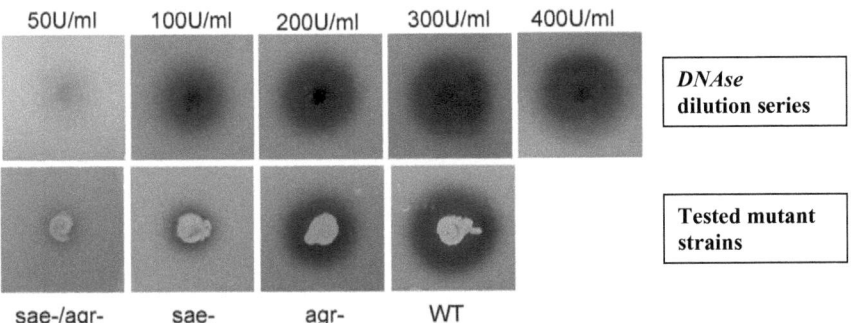

Figure 3-12: *DNAse* **production** *in vitro***:** The *DNAse* production was evaluated qualitatively via measuring the areola diameter size 1st in a *wild-type-*, 2nd in an *agr*$^-$-, 3rd in a *sae*$^-$- and 4th in an *agr*$^-$/*sae*$^-$ *SA* strain. Furthermore, to get an idea of approximately how much DNAse was produced; the areola diameter size of the different strains was compared to areola diameter sizes produced by a dilution series with 50U/ml, 100U/ml, 200U/ml, 300U/ml and 400U/ml (source: own data and picture).

3 – Experiments and results

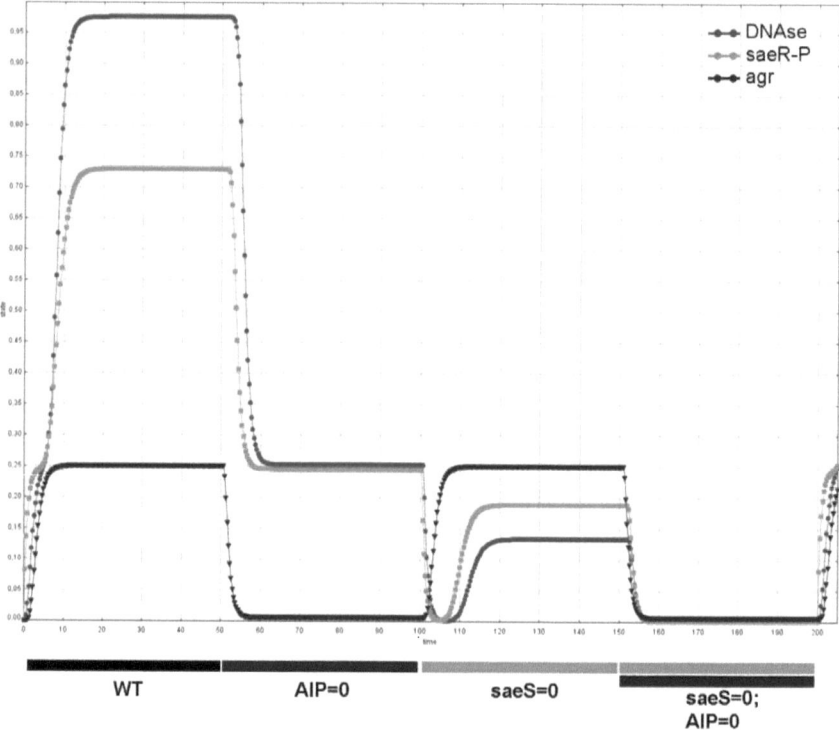

Figure 3-13: *DNAse* production *in silico*: The *DNAse* Production was simulated under 1st wild-type-, 2nd *agr*$^-$-, 3rd *sae*$^-$- and 4th *agr*$^-$/*sae*$^-$ conditions. Qualitatively the same *DNAse* production as in the *in vitro DNAse* experiments was found (see figure: 3-12) (source: own data and picture).

3.3.3.3 - Biofilm dissolution

After it was shown that the *DNAse* production in *sae*$^-$ strains is actually impaired the next goal was to find out if, by artificially adding *DNAse* to a *sae*$^-$ SA strain, the biofilm could be dissolved. This resembles a complementation experiment. Yet the complementation was not done genomically, because the *DNAse* deficit is not due to genomic reasons as well. The *DNAse*-locus is still intact in the *sae*$^-$ strains; it is just activated to a lesser amount, because it lacks the activating input of *sae*.

From figure 3-14 one can see that although the *DNAse* concentration is rising from left to right, the biofilm strength shows no consistency with that. With pure *DNAse* there are biofilm strength about as strong as with 300U/ml *DNAse* (for a summary of the different solutions and their composition, used for dilution of the biofilms see also table 3-6). This data leads to the conclusion, that not the high *DNAse* concentration, yet also not the high NaCl concentration is responsible for the weakening of the biofilms. Much more it seems that either the combination of both, the *DNAse* and the NaCl have a synergistic effect when it comes to dissolving a *SA* biofilm or that the DNAse-Buffer as a whole or one of its components leads to protection of the *SA* biofilms. Moreover a difference between the strains can be seen. The *agr⁻* biofilm seems to have under all five *DNAse* solutions the weakest biofilm, contrary to the *sae⁻* biofilms which seam to be the strongest when comparing the reactions to the five different *DNAse* concentrations. The double mutant *sae⁻/agr⁻ SA* strain seems to show a mixed phenotype when comparing the biofilm strength under the different *DNAse* concentrations. Under low *DNAse* concentrations this strain shows an *agr⁻* like phenotype, under high *DNAse* concentrations it shows a *sae⁻* like phenotype.

The next experiments, done to find out the cause for these very unexpected results, are summed up in figure 3-15. Here one can see that whether using 250U/ml or 500U/ml *DNAse* makes no difference. Furthermore one can see that again the *WT* biofilm can easily be dissolved by all the solutions. Moreover in the samples, where high NaCl concentrations were applied, the biofilm was widely dissolved. Higher H_2O concentrations also lead to weak biofilms except for the *agr⁻* strains where the biofilms weren't dissolved that much. High DNAse-Buffer concentrations on the other hand lead to weaker dissolution and thus to stronger biofilms. These results support the theory, that the DNAse-Buffer or one of its components protects the biofilm from being dissolved. Moreover from these results one can tell that the *agr⁻* biofilms have a strong resistance against being dissolved by H_2O. Yet this protective system against biofilm dissolution in H_2O is only active in the *agr⁻* biofilms and seems to be disturbed by NaCl. Because of this one can only find strong biofilms under high DNAse-Buffer concentrations and in the *agr⁻* biofilms under H_2O with low NaCl concentrations.

3 – Experiments and results

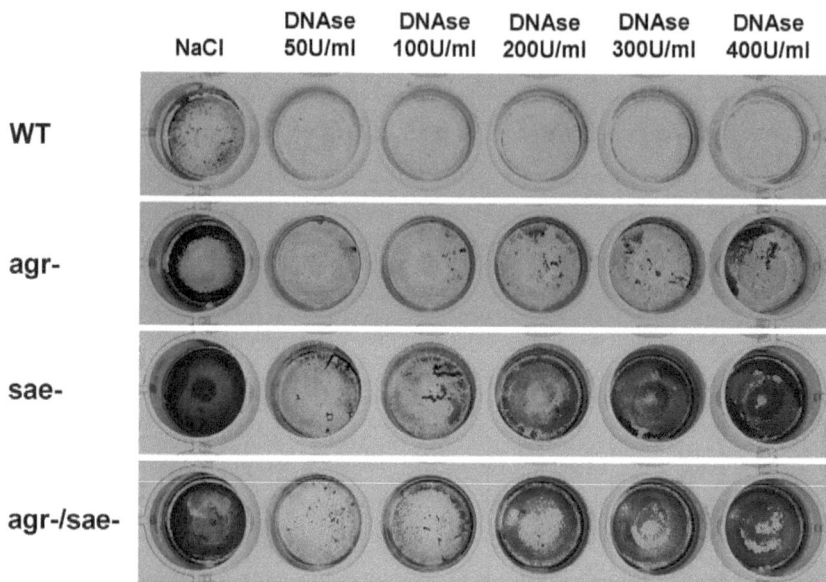

Figure 3-14: Effect of different *DNAse*/NaCl solutions on biofilms of different *SA* strains: In this figure one can see that, very unexpected, the strength of the biofilms is increasing with increasing *DNAse* concentrations. Yet, strains with pure NaCl show biofilms at least stronger than with 300U/ml *DNAse*. The *DNAse* from the stock is diluted in DNAse-Buffer at 1000U/ml. This *DNAse*/DNAse-Buffer solution was then mixed with 0.14M NaCl to get the different *DNAse* concentrations. Thus the amount of Buffer is increasing with the *DNAse* concentration, the NaCl concentration on the other hand is declining with *DNAse* concentration (source: own data and picture).

3 – Experiments and results

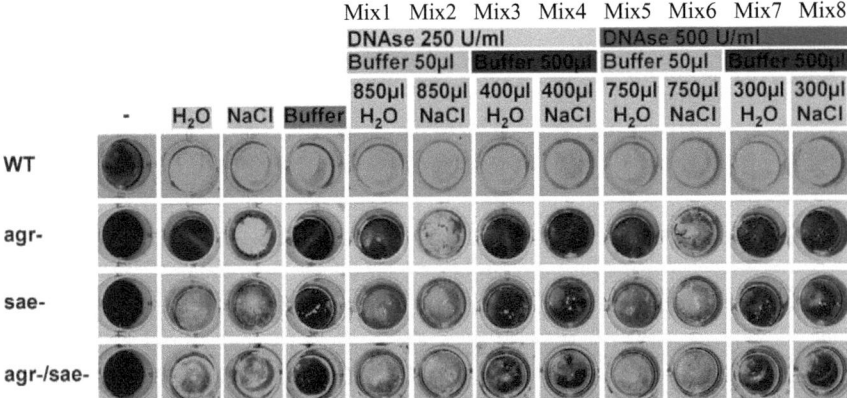

Figure 3-15: Effect of *DNAse*, DNAse-Buffer, H₂O and NaCl on different *SA* biofilms: Here the effect of *DNAse*, DNAse-Buffer, H₂O and NaCl on biofilms, build by *WT*, *agr⁻*, *sae⁻* and *sae⁻/agr⁻ SA* strains is shown. The substances were tested pure and in 8 different mixtures. *WT* strains can easily be dissolved by any of these solutions. The mutant strains are very prone to be dissolved by solutions with higher NaCl concentrations. Pure H₂O as well has strong biofilm dissolving capacities against *sae⁻*- and *sae⁻/agr⁻ SA* strains. Yet the biofilm of *agr⁻* strains aren't affected very much by H₂O. The Buffer seems to stabilize the biofilms, the *DNAse* in the concentrations used here doesn't seem to play an important role in dissolving the biofilms of all these mutant strains. Yet as shown in figure 3-14, strains with pure NaCl show biofilms at least as strong as with 300U/ml *DNAse* (source: own data and picture).

3 – Experiments and results

Table 3-6: Chemicals for dissolving *SA* biofilms and their concentrations in different solutions:

	H20	NaCl	DNAse-Buffer	*DNAse* 50U/ml
H_2O [ml]	1	1	1	1
DNAse [U/ml]	/	/	/	50
NaCl [mmol/ml]	/	0.14	0.082	0.13826
$MgCl_2$ [mmol/ml]	/	/	0.06	0.0018
$CaCl_2$ [mmol/ml]	/	/	0.01	0.0003
Tris [mmol/ml]	/	/	0.4	0.012

	DNAse 100U/ml	*DNAse* 200U/ml	*DNAse* 300U/ml	*DNAse* 400U/ml
H_2O [ml]	1	1	1	1
DNAse [U/ml]	100	200	300	400
NaCl [mmol/ml]	0.13652	0.13304	0.12956	0.12608
$MgCl_2$ [mmol/ml]	0.0036	0.0072	0.0108	0.0144
$CaCl_2$ [mmol/ml]	0.0006	0.0012	0.0018	0.0024
Tris [mmol/ml]	0.024	0.048	0.072	0.096

	Mixture 1	Mixture 2	Mixture 3	Mixture 4
H_2O [ml]	1	1	1	1
DNAse [U/ml]	250	250	250	250
NaCl [mmol/ml]	0.0181	0.1371	0.055	0.111
$MgCl_2$ [mmol/ml]	0.003	0.003	0.03	0.03
$CaCl_2$ [mmol/ml]	0.0005	0.0005	0.005	0.005
Tris [mmol/ml]	0.02	0.02	0.2	0.2

	Mixture 5	Mixture 6	Mixture 7	Mixture 8
H_2O [ml]	1	1	1	1
DNAse [U/ml]	500	500	500	500
NaCl [mmol/ml]	0.0321	0.1371	0.069	0.111
$MgCl_2$ [mmol/ml]	0.003	0.003	0.03	0.03
$CaCl_2$ [mmol/ml]	0.0005	0.0005	0.005	0.005
Tris [mmol/ml]	0.02	0.02	0.2	0.2

Here the exact concentrations of the different solutions, used for dissolving the *SA* biofilms are shown.

3.4 - Biofilm composition

For getting an even deeper insight in the composition of the biofilms, produced by agr^-, sae^- and agr^-/sae^- *SA* strains experiments were performed to qualitatively define the amount of nucleic acids, proteins and polysaccharides in biofilms of these mutants.

3.4.1 - DNA detection

Each of the three strains (agr^-; sae^-; agr^-/sae^-) was grown three times and then measured four times (see figure 3-16). Yet in all these experiments the differences between the gels were

3 – Experiments and results

stronger than the differences between the strains. Moreover the differences between the strains were not consistent. Thus one can say that no differences between these tree strains are findable and thus it needs to be proposed that qualitatively the amount of nucleic acids in the biofilms of these three strains is qualitatively the same.

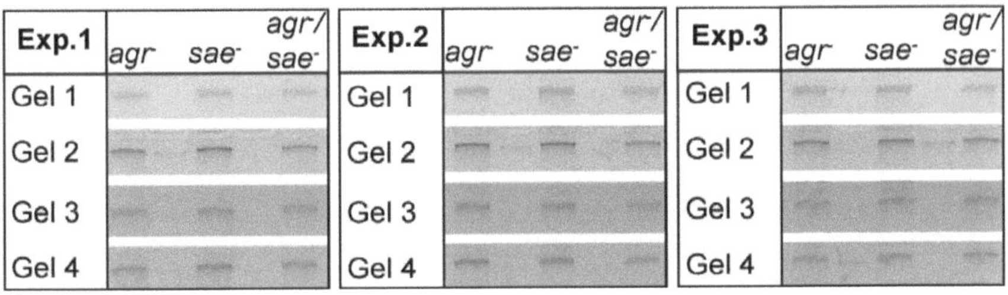

Exp.1	agr^-	sae^-	agr^-/sae^-	Exp.2	agr^-	sae^-	agr^-/sae^-	Exp.3	agr^-	sae^-	agr^-/sae^-
Gel 1				Gel 1				Gel 1			
Gel 2				Gel 2				Gel 2			
Gel 3				Gel 3				Gel 3			
Gel 4				Gel 4				Gel 4			

Figure 3-16: Amount of nucleic acids in different biofilms, qualitatively evaluated by blotting: In this figure one can see blots that show qualitatively the amount of nucleic acids in different biofilms. Biofilms of each strain were grown three times. The amount of nucleic acids in each biofilm was then measured four times in a gel. In all the experiments there is neither a big nor a consistent difference between the three strains (source: own data and picture).

3.4.2 - Protein detection

Each of the three *SA* strains (agr^-; sae^-; agr^-/sae^-) was grown three times and then measured four times (see table 3-7). The mean of these four measurements was calculated and the strains were ranked by their average OD at 595nm. In all three experiments sae^- *SA* strains showed the highest OD, followed by the agr^- strain. The weakest OD showed the agr^-/sae^- double mutant *SA* strain. Because in this assay the OD at 595nm strongly correlates with the amount of protein in the solution [10] it can be proposed that sae^- *SA* strains produce the highest amounts of protein in their biofilms, followed by the agr^- *SA* strain. The agr^-/sae^- double mutant *SA* strain showed the weakest amount of protein in its biofilms.

3 – Experiments and results

Table 3-7: OD's measured in the different biofilms by Bradford-Protein-Assays:

	agr^-	sae^-	agr^-/sae^-
Average OD Exp. 1	0.083	0.085	0.073
Rank Exp. 1	2	1	3
Average OD Exp. 2	0.088	0.101	0.075
Rank Exp. 2	2	1	3
Average OD Exp. 3	0.095	0.110	0.093
Rank Exp. 3	2	1	3
Average OD Exp. 1-3	0.089	0.099	0.080
Rank Exp. 1-3	2	1	3

In this table the OD's (at 595nm against H_2O) of the Bradford-Protein-Assays that were done with the different biofilms are shown. Moreover the ranks of the different *SA* Strains are shown. Biofilms of each strain were grown three times. Each biofilm was then measured four times; the average values of these measurements are shown here. In all three experiments sae^- *SA* strains showed the strongest protein production, followed by the agr^- *SA* strain. The agr^-/sae^- double mutant *SA* strain showed the weakest protein production.

3.4.3 - Polysaccharide detection

Each of the three *SA* strain (agr^-; sae^-; agr^-/sae^-) was grown three times and then measured four times (see table 3-8). The mean of these four measurements was calculated and the strains were ranked by their average OD at 480nm. In all three experiments agr^- *SA* strains showed the highest OD, followed by the sae^- *SA* strain. The weakest OD showed the agr^-/sae^- double mutant *SA* strain. Because in this assay the OD at 480nm strongly correlates with the amount of polysaccharides in the solution [27] it can be proposed that agr^- *SA* strains produce the highest amounts of polysaccharides in their biofilms, followed by the sae^- *SA* strain. The agr^-/sae^- double mutant *SA* strain showed the weakest amount of polysaccharides in its biofilms.

3 – Experiments and results

Table 3-8: OD's measured in the different biofilms by a Polysaccharide-Assay:

	agr^-	sae^-	agr^-/sae^-
Average OD Exp. 1	0.130	0.080	0.067
Rank Exp. 1	1	2	3
Average OD Exp. 2	0.177	0.111	0.077
Rank Êxp. 2	1	2	3
Average OD Exp. 3	0.139	0.100	0.064
Rank Exp. 3	1	2	3
Average OD Exp. 1-3	0.149	0.097	0.069
Rank Exp. 1-3	1	2	3

In this table the OD's (at 480nm against H_2O) of the Polysaccharide-Assays that were done with the different biofilms are shown. Moreover the ranks of the different *SA* strains are shown. Biofilms of each strain were grown three times. Each biofilm was then measured four times; the average values of these measurements are shown here. In all three experiments agr^- *SA* strains showed the strongest polysaccharide production, followed by the sae^- *SA* strain. The agr^-/sae^- *SA* double mutant strain showed the weakest polysaccharide production.

From these experiments one gets a hint on the composition of biofilms, produced by agr^-, sae^- and agr^-/sae^- mutant *SA* strains. Agr^- biofilms, compared with the two other tested *SA* strains, have the largest amounts of polysaccharides; comprise a protein amount which lies between that of the sae^- and agr^-/sae^- biofilms and has approximately the same amount of nucleic acids like the two other *SA* strains. Sae^- biofilms, compared with the two other tested *SA* strains, have the largest amounts of proteins; comprise a polysaccharide amount which lies between that of the agr^- and agr^-/sae^- biofilms and has approximately the same amount of nucleic acids like the two other *SA* strains. Agr^-/sae^- biofilms, compared with the two other tested *SA* strains, have the weakest amounts of polysaccharides and proteins, yet they also comprise approximately the same amount of nucleic acids like the two other *SA* strains.

Table 3-9: Overview across the composition of different biofilms:

	agr^-	sae^-	agr^-/sae^-
nucleic acids	+	+	+
proteins	++	+++	+
polysaccharides	+++	++	+

Here the composition of agr^-, sae^- and agr^-/sae^- biofilms regarding nucleic acids, proteins and polysaccharides are shown.

4 – Discussion

4 - Discussion

4.1 - Discussing the results

In this concrete example the *QS* model that was build is a network that focuses on several central regulatory nodes like for example the two component systems *agr, sae* or *arl* and many other different crucial nodes and signalling cascades like *SigB, Rot* or *Sar*. This network model provides plenty of detail such as for example the, in detail modeled, regulation of the *agr*-locus itself or all the nodes and their connections included in the periphery. Thus this model is obviously still simplified concerning the exact interaction of the nodes on a more genetic, molecular level. For example no difference was assumed, whether an inhibiting connection influences the transcription, the translation or the performance of the final gene product. In this case it is only important that the effect of the particular node is affected, regardless in which way. Moreover the network model works independent of exact kinetic data. This network model not only provides a basis for simulating *QS,* yet in addition to the own experiments and the comparison to published data; it helps to better elucidate the functions and interactions around the central nodes of regulation. The network was validated by comparing its output to micro array data; off course analyzing only the gene expression data doesn't fully cover all eventualities the network model covers. For example there could be an interaction between two nodes solely on the protein level, which one then would miss. Yet also when analyzing direct protein-protein interactions, such as for example protein phosphorylation, this couldn't be done in such a broad approach and the functionality and efficiency of the proteins are still not elucidated. Thus, the aim was to find an easy and effective method, providing in a broad approach a good secondary indicator for the node activities of a large amount of nodes, incorporated in this network model. This was found in the microarrays that were analyzed and the northern blots that were conducted. Furthermore the network was successfully used to make predictions about the outcome of Northern blots and biofilm adherence assays. This leads to the conclusion that this network could be used for quick and easy testing of predictions around the *agr*- and the *sae*-locus, as well as the *QS* of *SA*. Knock-out mutants for example could first be tested *in silico* before creating them costly. Moreover the interaction of different nodes can be seen very easily and thus this network can be used for planning which node needs to be knocked-out or altered by drugs, to lead to a specific result. Due to its segmental construction and as easy to handle freeware was used this

4 – Discussion

network could easily be changed according to the current knowledge, adapted to different purposes or just extended.

Many of the nodes, embedded in this *SA QS* network, are known to play an important role also in other staphylococci [90]. The *agr*-locus for example is not only existent in *SA*, yet it also plays an important role in *QS* and virulence in *Staphylococcus epidermidis* (*SE*). In *SE* the gene structure and the sequence of the *agr* locus is quite similar to that of *SA* and may hence play a comparable role as in *SA* [112]. The s*ae* two component system on the other hand for example also plays an important role in *Staphylococcus carnosus* [90] and *SE* [37].

Besides providing a comprehensive network around the *agr*-locus and the biofilm regulation of *SA*, this work also shows that the *sae*-locus might be of greater importance for the biofilm regulation than thought until now. Biofilms are formed stronger under *sae*$^-$ conditions and are impaired again when *sae* is complemented. By creating Venn-diagrams it was shown that nucleases, as well as extracellular and plasma binding proteins (*SA0744*), which are under control of the *sae*-locus, might be an important regulatory link between the *sae*-locus and the biofilm formation ability. To evaluate this hypothesis, experiments were done to gain more knowledge about the biofilm composition and its reaction to nucleases. Therefore the polysaccharide-, the protein- and the nucleic acid concentration, in the biofilms of the different strains, was evaluated. Moreover the *DNAse* production of the different strains was qualitatively evaluated and experiments were done to dissolve the biofilm with *DNAse*. Yet these *DNAse* experiments show how sensitive the used *DNAses* and also the biofilms are against different chemicals. This made it hard to tell which reactions are clearly due to the different chemicals. For example it was shown for *SE* that biofilm formation can be enhanced by Mg_2^+ and inhibited by EDTA [31], yet $MgCl_2$ is an essencial component of the DNAse-Buffer, needed by the *DNAse* to work properly. Therefore the biofilm dissolution assay was repeated with mixtures in which we hold different ingredients of the solution at a constant concentration to distinguish which ingredient could be responsible for the *DNAse* independent biofilm dissolution effect. We couldn't find differences in the biofilm dissolution strength between the *DNAse* concentrations. The biofilms treated with higher concentrations of the buffer ($MgCl_2$, $CaCl2$ and Tris) showed more stable biofilms, which agrees nicely with the results known from *SE* [31]. Mixtures with higher NaCl and with low buffer concentrations dissolved the biofilm properly. Only the *agr*$^-$ biofilms seem not to be dependant on the biofilm strengthening effect of on of the buffer components. In summary the results show that the biofilm can be strengthened by one of the buffer components, yet which of the components is responsible for this effect still needs to be determined. Moreover these results show that NaCl is either a factor leading to biofilm dissolution or has at least an effect on the *agr⁻* biofilm, interfering with the system, making this biofilm independent from the biofilm strengthening effect of the buffer.

In the experiments, concerning the biofilm composition, this work shows that the *agr⁻* biofilms have the highest polysaccharide concentration and that *sae⁻* biofilms have the highest protein concentration. This supports the hypothesis that proteins are important for the *sae* controlled (inhibited) biofilm formation, suggesting that the proteins are down-regulated by *sae* in parallel to the biofilm forming ability. A problem in the experimental design here is that the biofilm amount, in the tested strains, is not quantitatively assignable, yet we qualitatively showed, in the biofilm adherence assay, the biofilm formation strength to be about equal in the three tested strains. When looking at the results it is obvious that the differences between the strains cannot just be due to a different strength in biofilm formation, because first of all there was no relevant difference, regarding the biofilm strength as can be seen in figure 3-9. Moreover when comparing the results for the *agr⁻* and *sae⁻* strains, regarding the polysaccarid and the protein concentrations in the biofilms one finds that in the *sae⁻* strain the proteine concentration is higher than in the *agr⁻* strain and for the polysaccharide concentration vice versa. This couldn't be found if one of the two strains would produce a clearly stronger biofilm. Only the *agr⁻/sae⁻* strain shows in all the tested biofilm componets the lowest concentration which could be due to a weaker overall biofilm formation, yet which couldn't really be supported by the biofilm adherence assay.

4.2 - Related work

4.2.1 - *QS* simulations around the *agr*-locus of *SA*

SA, their biofilms and thus also the *QS* of *SA* are of great importance in medicine and research (see chapter 1.1.1 - 1.1.3). As a result this not the first thesis, concerning the simulate of the *QS* around the *agr*-locus of *SA*.

Jabbari et al. [51] for example used a mathematical modelling approach with which they simulated solely the *agr*-locus of *SA*. Therefore a set of differential equations was created (see figure 4-1) with which the processes around the *agr*-locus can be described properly. A few assumptions were made for setting up this differential equation system. First of all it was assumed that all products are subject to a natural degradation and dilution process. For example when bacteria are undergoing binary fission, without protein production the concentration of the products are diluted. Moreover it was assumed that receptor (*AgrC*)

4 – Discussion

bound *AIP* can unbind spontaneously with a certain possibility, that *AgrA-P* can be dephosphorylated to *AgrA* by housekeeping phosphatases with certain rate and that all reactions, for example phosphorylation, don't take time themselves. For the initial conditions Jabbari et al. [51] used a steady state without any *AIP* production, thus the product levels are controlled just by the basical protein expression on the one hand and the natural degradation and dilution processes on the other hand. Values for these processes were, as far as possible, based on experimental evidence. Using these conditions allows simulating how a large bacteria population shifts from down- to up-regulated due to increasing *AIP* levels.

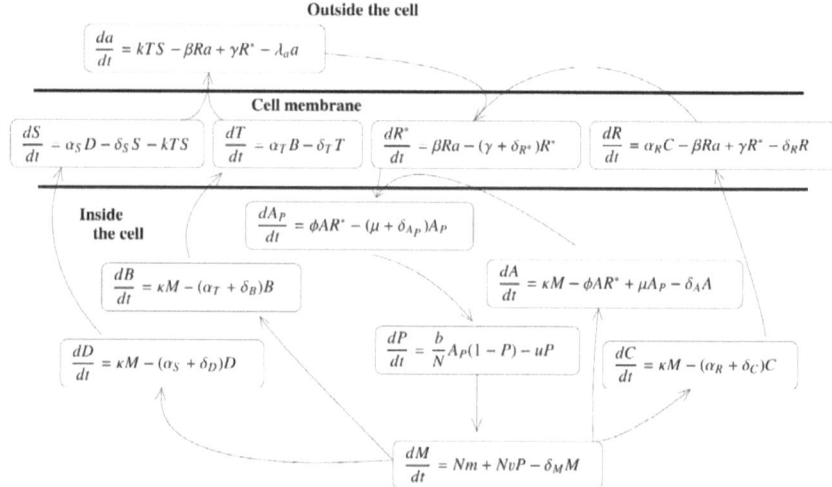

Figure 4-1: Mathmatical simulation of the *agr*-locus: This scheme shows a set of differential equations to simulate the processes around the *agr*-locus mathematically [51] (for more details see Jabbari et al. 2009 [51]).

Another example is the mathematical modelling approach, used by Gustaffson et al. [43]. This approach also includes the *agr*-locus, but in contrast to the simulation of Jabbari et al. [51], here the influence of *SarA* was also taken into account and the *AIP* concentration was treated as a parameter of the model and not as a result of it. Thus not the full *SA QS* feedback circuit around the *agr*-locus is included here. The reaction of the *agr*-locus, meaning the intracellular *RNAIII* concentration, was simulated as a function of the *AIP* concentration in a homogeneous, none growing, bacterial population. This was done disregarding the *AIP* production; to be able to analyze the *agr* reaction to fixed *AIP* concentrations, yet this also disrupts the *AIP* feedback loop. The simulation itself was build up of a set of ordinary

4 – Discussion

differential equations, following fundamental kinetic principles, akin to those described for Jabbari et al. [51].

A slightly different simulation of the *agr*-locus was done by Koerber et al. [61]. Besides living planctonic or in a biofilm, *SA* is also able to live intracellular in endosomes. Therefore *SA* infiltrates non professional phagocytes through interaction between its fibronectin binding proteins and the host cell fibronectin. Typically only one bacterium is then enclosed in such an endosome and escapes from it again into the cytoplasm of the host cell to reproduce. This usually leads to the death of the cell. This mechanism of course can be assumed to play an important role for the antibiotic and immune system resistance and thus for the persistence of *SA* infections. The mechanism of escaping out of the endosome seems to be, like the *QS*, regulated by the *agr*-locus. *SA* shows increased *agr* stimulation prior to endosome escape and *agr* defective bacteria are not able to replicate intracellular anymore. It is thus hypothesized, that here the same mechanisms are at work like in *agr* mediated biofilm dispersal. The *AIP* concentration rises, in this case due to no dilution instead of increased cell density, and hence the *agr*-locus is up-regulated, resulting in increased *AIP* production again. This leads to a positive feedback mechanism, yet also to increased production of lysines and toxins, normally mediating biofilm dispersal, that are able to dissolve the endosome and release the bacterium into the cytoplasma. This means that here dilution sensing or compartment sensing, instead of cell density sensing takes place and needs to be simulated. Therefore Koerber et al. [61] developed a stochastic model, regarding the *agr*-locus, in which the bacterium stochastically is either down- or up- regulated; because with just one bacterium in an endosome a continuum between these two states cannot be simulated. In this model the probability of the bacteria, to be either in the down- or up- regulated state and the probability to switch from the one state to the other, is calculated also using differential equations.

Thus in conclusion, with the network model this is not the first thesis in which the *QS* of *SA is simulated*, yet it is the first to simulate it as a Boolean network and with including so many nodes. Moreover in this thesis it is proven that the network qualitatively reflects the *in vitro* processes and by using easy to handle freeware, for building and simulating this network, it is easy to handle and extendable. Thus the network is the first one around the *QS* of *SA* which is easy to handle, easy extendable, comprehensive and validated.

4 - Discussion

4.2.2 - SQUAD simulation of survival and apoptosis in liver cells (Philippi et al. 2009)

SQUAD was also used by Philippi et al. [83] to simulate a network. In this case the simulation modelled the system states of liver cells, either survival or apoptosis, which they adopt in response to viral infections. Therefore also an extensive literature research, including different databases, was done; to get first of all a comprehensive network (74 nodes, 108 edges) of proteins, involved in the apoptosis signalling of hepatocytes around the *Fas* ligant (*FasL*) mediated apoptosis. Also proteins for crosstalk were included (see fgure 4-2). Four steady states were found in this network. Steady state A1 correlates to cells in suspension, it represents the mitochondrial/intrinsic pathway. Here all caspases are active, *Bcl-2* is inactive and *cytochrome c* is released into the cytoplasm. Steady state A2 represents the extrinsic pathway, like in cells grown on collagen. *Bcl-2* is active but not interfering with the extrinsic pathway. State A3 and state A4 represent stable, non apoptotic states in which *AKT* and *NFκB* are up-regulated. The disparity between these two steady states is the up- or down-regulation of *Bcl-2* and *Stat3*.

In addition to the basic network, extended and modified models representing different viral infections were constructed.

For validation of the model it was compared to data from experiments about kinetics of *caspase* activation and *cytochrome c* release in wild type and *Bid* knock-out cells, grown on different substrates.

As an example for medical applications, cytomegalovirus proteins *M36* and *M45* were added to the basic model. Both proteins were found to be responsible for blocking the apoptotic pathways, just leaving the pathways leading to survival. *M36* blocks the mitochondrial apoptosis pathway via *Bcl2*; *M45* blocks the apoptotic pathway via *RIP1*, a crucial protein kinase of the death receptor complex. These tests again produced output behaviour that well agrees with experimental data.

This work is a good example for how SQUAD simulations can be used to create comprehensive networks to get new insights in complex pathways and signalling cascades.

4 – Discussion

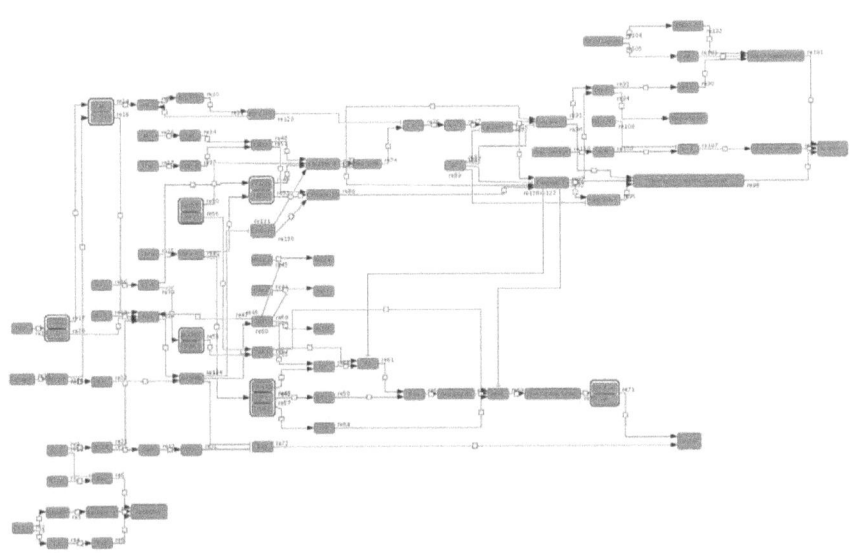

Figure 4-2: Schematic view of major apoptosis pathways in mammalian cells: Comprehensive network showing proteins involved in liver cell apoptosis, including their, either up- or down-regulating, interactions. Also different nodes around key pathways of apoptosis in hepatocytes, as well as a number of proteins implicated in crosstalk are shown (source: Philippi et al. 2009 [83]).

4.2.3 - A *QS* regulated trade off in biofilms (Bassler et al. 2011)

For testing, whether there is actually a *QS* based trade off between on the one hand *EPS* production and thus decision for a life in a biofilm and on the other side *EPS* repression to leave an existing biofilm or to avoid getting part of one, Bassler et al. [72] created *V. cholerae* (*VC*) *ΔflaA; ΔhapR* double mutants.

Biofilm growth and thus *EPS* production is initiated in *VC* after adhering to surfaces, normally going along with dropping its flagella. Of great importance for the flagella is the gene locus *flaA* which encodes the flagella core protein Flagellin. Moreover *QS* around *hapR*, the master *QS* regulator is used by *VC* to up-regulate *EPS* production at a low cell density level and also to down-regulate *EPS* expression at high cell density levels. This *ΔflaA; ΔhapR* double mutant thus produces *EPS* on a basical level and is thus in this text labelled EPS^+.

In addition to get an EPS^- strain, with the least genetic difference to the EPS^+ mutant a *ΔflaA; ΔhapR; ΔvpsL* triple mutant was constructed that never produces *EPS* because *vpsL* is essential for the *EPS* biosynthesis. This triple mutant was thus called EPS^-.

4 – Discussion

The growth rate of the EPS^- strain was 25% higher than that of the EPS^+ strain. This shows very clear that the EPS^+ strains are impaired in their growth rate, most likely because of taking away resources from the biomass production and redirecting them into the synthesis of EPS. Yet EPS production is also supposed to come with a competitive advantage. To figure out whether such a benefit is specific in the biofilm surroundings EPS^+ and EPS^- strains were inoculated in coculture and in monoculture. The EPS^+ strain accumulateded, per unit area of substratum, more biovolume than the EPS^- strain when growing these two strains in a biofilm monoculture. Furthermore when growing these two strains together, what means growing them in biofilm coculture at a ratio 1:1, the growth of the EPS^- strain is impaired by more than 80% whereas the EPS^+ strains biovolume accumulation remains unaffected. All this suggests that the EPS-production is a fitness advantage in biofilm environments (see also fgure 4-3).

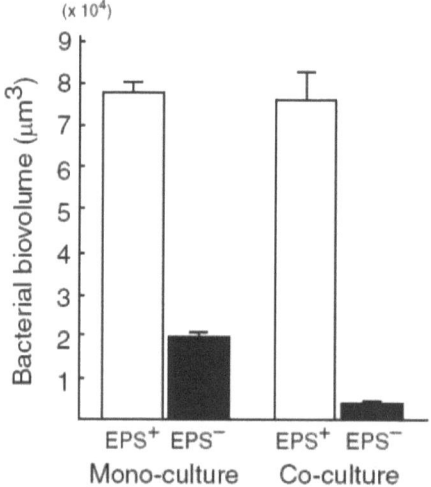

Figure 4-3: Growth of EPS^+ and EPS^- strains in biofilm culture: In biofilm environment EPS producing strains (white bars) show stronger growth than EPS^- strains (black bars). When grown in co-culture EPS^+ strains are not impaired in their growth in contrast the biovolume accumulation is reduced in EPS^- strains (source: Bassler et al. 2011 [72]).

With the biofilm structures they rendered from the confocal micrograph stacks Bassler et al. were able to show that the growth impairment of the EPS^- strains in coculture is at least partly just due to physical displacement of the EPS^- cells -which are confined to the substratum- by EPS^+ cells able to divide and grow into three dimensional clusters (see also figure 4-4).

4 – Discussion

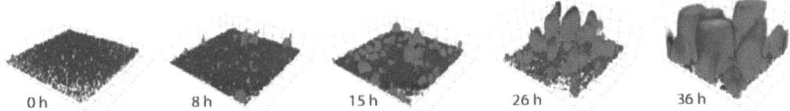

Figure 4-4: Biofilm structures rendered from confocal micrograph stacks: EPS^+ cells (light gray) grown in a biofilm together with EPS^- cells (dark gray). Gradually the EPS^+ cells crowd out the EPS^- cells. After 36h the summits of some EPS^+ cell clusters are flat, because they touch the ceiling of their growing chamber. Grid boxes were 11µm on each side (source: Bassler et al. 2011 [72]).

These results suggest that EPS^- strains have a fitness advantage in liquid planctonic environments where simply the growth rate counts. In biofilm environments on the other hand where EPS is essential for the stability of the biofilm EPS producing strains have a fitness advantage.

Yet beyond simple growth rate, organisms and thus also bacteria need to move to new resource spots, as soon as the old spots are destroyed or depleted. Several bacterial biofilms are known to disperse, most likely regulated by QS mechanisms, as a reaction to specific environmental cues. Bassler et al. thus took effort to figure out the impact of EPS-production on this dispersal ability in VC. Therefore EPS^+ and EPS^- strains were inoculated in microfluidic chambers and after 20h and 46h the biofilm chamber's effluent was examined; with the result that EPS^+ cells were sparsely represented at both time points (see also fgure 4-5).

4 – Discussion

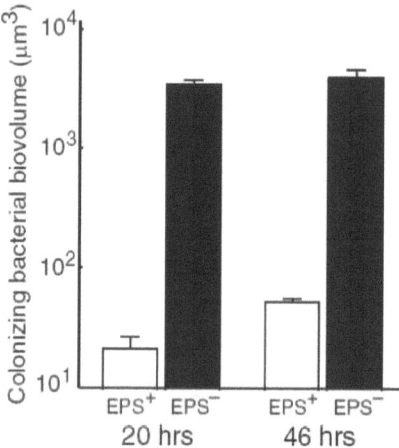

Figure 4-5: Colonizing bacterial biovolume after dispersal from biofilms: This figure shows the biovolumes, in the freshly formed monolayers of EPS^- cells (black bars) and of EPS^+ cells (white bars) when diverting efflux from the chambers containing the growing biofilms to fresh chambers at 20h and 46h. In the newly formed monolayers from both time points EPS^- cells are predominant suggesting better dispersal abilities in this strain (source: Bassler et al. 2011 [72]).

This system is an approved paradigm in ecology, known as the the competition-colonization trade-off. Here the QS regulated EPS-production is advantageous in one scenario, yet disadvantageous in the other scenario. EPS-production is advantageous -and thus should be up-regulated by QS- in biofilms and stable environmental conditions, with no need for dispersal, for example when the resources in the colonized patches are vast and patches are not often destroyed. Yet in planctonic growth conditions and in unstable environments, where dispersal is inevitable, EPS-production is disadvantageous and should thus be down-regulated by QS.

With this work Bassler et al. demonstrates the great impact QS can have in bacteria and how fine tuned and balanced these regulatory mechanisms are. Moreover this work shows, like very often seen in swarm or collective intelligence, that such relatively simple processes could possibly lead to decision making processes of such great importance when looking at it on a bigger scale.

4 – Discussion

4.2.4 - Global gene expression in *SA* biofilms
(Beenken et al. 2004)

For identification of gene loci that could possibly be relevant for the formation of biofilms or the adaptive reaction of *SA*, required for the persistence within a biofilm, Beenken et al. [6] isolated all the cellular RNA from the *SA* strain *UAMS-1*, grown in a biofilm, as well as from both, exponential- and stationary-phase planctonic cultures of *UAMS-1*. The gene expression was evaluated using a special, customized Affymetrix GeneChip that represents the genomic analogue of six *SA* strains (COL, NCTC 8325, N315, Mu50, MSSA-476 and EMRSA-16 [strain 252]). Afterwards the RNA expression in planctonically grown bacteria was compared to the RNA expression in bacteria grown in a biofilm. In these experiments 48 genes were found to be up-regulated at least by the twofold in bacteria, grown in biofilms, compared to planctonically grown bacteria. Moreover 84 genes were identified, being repressed at least by factor two in bacteria, grown in biofilms, compared to planctonically grown bacteria. To verify these results real-time PCR experiments were done to evaluate the relative expression levels of selected genes. The gene expression patterns, observed with real-time PCR, were concordant with the microarray experiments. Yet the PCR results suggest that the microarray data may lead to underestimation of the actual differences.

Moreover it was found by Beenken et al. that *SarA* defective strains had impairment in biofilm formation and thus they suggested this locus to play a major role in biofilm formation by changing the expression of downstream genes. Beenken et al. found 27 genes under the control of *sarA* and in addition variably expressed in planctonically grown bacteria and bacteria grown in biofilms. Four of these 27 genes are found to be up-regulated in biofilms and also by *sarA* (*sdhB*, *carA*, a hypothetical protein and an unidentified ORF which has a similarity to an analogon of the major histocompatibility complex [MHC] class II). Further eight of these 27 genes were found to be down-regulated by *sarA* and in biofilms (*arc*, *phoP*, *pbp3*, *nuc*, *ndhG*, *spa*, and two hypothetical proteins). The remaining 15 genes were deviant regulated in biofilms and by *sarA*.

With this work Beenken et al. provides a lot of data about the gene expression in *SA* biofilms and thus lays a cornerstone for work, revealing remaining secrets of *SA* biofilms, like further *SA* biofilm analysis efforts.

4 - Discussion

4.3 - Relevance of the work presented in this thesis

QS is an intriguing and well studied feature of bacterial adaption. This includes first simulation and modelling efforts [93; 114; 51]. *QS* can be seen as a very basic model for swarm intelligence (which is an emergent and collective intelligence in groups of simple agents [9]) and as a model for basic decision making processes, where the regulatory networks of the individual bacterial cells are the basic entities for the emergent and collective behaviour. This network is a major regulator of *SA* to change between the planctonic and the biofilm state of living [120]. Because this change comes along with massive changes in pathogenicity and resistance to the host's immune system and antibiotics, controlling this state change can potentially be used for the treatment of infections, caused by *SA* [8]. By realizing the broad influence of *QS* in the infective behaviour of many bacterial species, one can see an auspicious, on-going effort in developing small molecules that target *QS*. Anti-biofilm agents, such as *QS*-supportive drugs, may be beneficial in order to facilitate bacterial clearance by the immune system and/or antibiotics. Yet the anti-*QS* drugs, known as quorum quenchers [99], on the other hand could prevent *SA* from changing from the relative offenceless, biofilm building to the noxious, invasive phenotype and thus prevent *SA* from expressing noxious factors, like for example the toxic shock syndrome toxin (TSST) and also from invading healthy tissue. All this could potentially be reached by interfering for example in the *QS* of *SA*. Anyhow, all factors leading to stronger biofilms or expression up-regulation of virulence factors are potential targets for new, innovative, anti-staphylococcal agents. Besides the medical implications, controlling biofilm formation behaviour and *QS* in *SA* and in bacteria in general, has also a vast impact in other scientific disciplines. There are for example the building and construction of micro structures in self assembling systems. Another example is exploiting the swarm intelligence, like the decision making capability of the *QS* process, to incorporate it in elementary biological computing systems, such as logic operators.

Thus this work not only provides a powerful tool for further investigations, concerning *SA*, its biofilm, its *QS* and the *agr*-locus, but also shows the importance of *sae* for the biofilm formation ability. All this brings us a step further in fighting *SA*, one of the most widespread and most important commensal bacteria [65]. Thus this work is of great importance in medicine. Yet this work is also of great importance in Biology. First, concerning *QS* mechanisms, for example when trying to understand principle mechanism in *QS*, this network could be very helpful. Second, concerning the *agr*-locus, here also the network could be very helpful. Third, concerning biofilms, their formation and their composition, here also the network, yet also the used methods, and the gained results in evaluating the biofilm composition, could be very helpful.

4.4 - Future work

4.4.1 - Using the network

In further experiments it would be obviously of great interest to use this network to gain more knowledge about all the included nodes. For example the impact of *arl* or *sar* on the biofilm formation ability, but also on other QS regulated mechanisms, like the reaction to stressors, such as nutrient depletion, could be evaluated. Moreover this network could be used to get a deeper insight in how all the included nodes interact, just by playing around with the activation of the different nodes or by systematically down- and up-regulating different specific nodes to predefined levels and then evaluating the results.

Besides just using the existing network a big effort should be spend on keeping this network up to date and extending it with new nodes and edges. Thus finally one would get a really comprehensive network, not only around the a*gr*-locus, the QS and its impact on the biofilm forming ability of SA, but covering all the regulations and interactions taking place in SA. Finally, yet nowadays a bit utopistic, one could then be able to simulate one hole SA bacterium and maybe even a whole SA colony.

4.4.2 - More detailed examination of biofilm composition.

In this work the biofilm composition of different mutant strains was examined and compared. Qualitatively the amount of nucleic acids, proteins and polysaccharides in these biofilms was investigated. Regarding the nucleic acids there were no concentration differences findable between the different biofilms. Regarding proteins and polysaccharides there are qualitative differences in their concentration between the biofilms of the different mutant strains. In further experiments it would thus be of great importance and interest to improve the experimental methods, to enable us to first of all find smaller differences in the amounts of these three substances and thus for example be able to differentiate between the concentrations of the nucleic acids in the biofilms of the tested mutant strains. Yet this would give us also the possibility to differentiate the nucleic acids, polysaccharides and proteins and moreover to examine other substances, maybe related to biofilms, such as lipids. Yet besides all this it would also be of great importance and interest to go a step further and also test the biofilms of other mutant strains, in which other regulatory loci are up- or down-regulated or

4 – Discussion

even knocked-out. All this would help to get a more detailed knowledge about the composition and also the differences between existing biofilms, not only produced by *SA* but also by other microorganism. Finally this could lead us a step further in knowing how to regulate, fight and use these biofilms.

4.4.3 - Solving the DNAse dissolution problems

For this work experiments were done to evaluate the reaction of different biofilms to *DNAse*. Therefore the *DNAse* production of the different strains was qualitatively evaluated and experiments were done to dissolve the biofilm with *DNAse*. Yet these *DNAse* experiments provided results, not sufficiently explainable just by the effect of the *DNAse* itself and thus showed how sensitive the used *DNAses* and also the biofilms are against different chemicals and environmental conditions. This made it hard to clearly tell, which reactions are due to the different chemicals. In further experiments one goal should be to separate the effect of the *DNAse* from the effects of the other chemicals, which is not that easy, because for example the effectivity of the *DNAse* is partly also dependant on these chemicals. Yet when one could really differentiate between the effects of the different chemicals, this would allow us to get greater influence on biofilms and again would lead us a step further in knowing how to regulate, fight and use these biofilms.

Besides investigating the influence of *DNAse* on the biofilm formation ability, *SA0744* (*extracellular ECM and plasma binding protein*) the second gene found in the Venn-diagrams should not be disregarded. One goal in future work should thus be to implement this gene as a further node in the network and to reveal the importance this node has for the whole network and the impact it has on the *QS* and biofilm forming ability *of SA*.

5 - Conclusion

As shown in chapter 4.2.1 there are already simulations regarding the *agr*-locus and the *QS* of *SA*, yet in this thesis, to my knowledge, the first Boolean network around the *agr*-locus is created and presented, including many different two component systems, such as *arl* and *sae* and other different important nodes and signalling cascades, such as *SigB, Rot* or *Sar*.

Another advantage of this network and this kind of simulation, compared to the other already existing networks and simulations is, that it is easy to work with and all the software that was used is freeware. The network can easily be extended with new nodes or regulatory circuits and adapted to new findings and insights. Moreover, in this network all available data about nodes and their interactions was included. This could be just the proven knowledge that in a specific knock out strain a special node is up- or down-regulated or the evidence for a precise interaction process, like for example the impact of a node on the transcription or translation of another node. The network model is comprehensive and reflects the whole knowledge about the *agr*-locus and the *QS* of *SA* available today and thus can be assumed to provide the best simulation results, bringing the *in silico* results, on a sufficiently detailed level, as close to the real world as possible nowadays.

This *SA QS* network presented here has two different steady states, one representing an invasive, toxic phenotype, the other one representing a biofilm producing phenotype. This network was validated by comparing it with Northern blots and microarrays of previous publications. Furthermore *in silico* predictions were made about the *QS*, the reaction of different nodes and the biofilm building ability of different mutant strains. These predictions were compared to *in vitro* experiments, such as Northern blots and biofilm adherence assays. In these experiments the predictions were found to be confirmed. Moreover, as far as I know, in this thesis for the first time it is shown *in vitro* and *in silico* that *sae* has a strong influence on the biofilm building ability of *SA*. When *sae* was knocked-out *in vitro* or *in silico* the biofilm building ability of *SA* was increased. By complementation of *sae* the influence of *sae* on the biofilm building ability was confirmed. Thus this network simulation was not only validated against existing data (micro array) and tested if it yields the right predictions for the experiments (Northern blots of knock out mutants), yet it also suggested that *sae* shuld play an important role in biofilm formation. This was proven by in vitro experiments (biofilm adherence assay).

5 – Conclusion

In addition a hypothesis is provided how *sae* could influence the biofilm formation so strongly. To prove this hypothesis first steps were taken in analyzing the biofilm composition and the impact of nucleases on different biofilms. The experiments showed clear differences in the composition of the agr^- and the sae^- biofilms. Thus in summary in this thesis the importance of the *sae*-locus for the biofilm formation is shown as well as the difference between a sae^- and an agr^- biofilm, regarding composition, yet also stability and resistance to dissolution.

Biofilms have a strong effect on survival and virulence of bacteria [26] thus the impact of *sae*, newly found and described here, could represent a first step for the invention and development of new drugs. These drugs could help to fight *SA* infections and also prevent *SA* from colonisation of, for example catheters. Here they very often form biofilms [8], in which they are harder to eradicate, because they are better protected from drugs, including recent antibiotics [8].

The established network and simulation allows studying *QS* and biofilm formation in *SA* and is made publically available. With the *in silico* network, shown in the results, that agrees qualitatively well to the transcriptome data and additional experiments, different knock out mutants, regarding the nodes of the network, are fast and easily simulated and compared to experimental data, such as gene expression data and experiments. Moreover, one gets predictions, well supported by the experimental data shown in this thesis. A basic and specific regulatory role of *agr*, *sae* and *arl* is delineated in this network model with extensive support from the obtained experimental data. Without this network model, most likely a lot more time of expensive experimental work would be needed. Though all *in silico* predictions need to be verified by *in vitro* experiments the model saves a lot of money and time.

6 - Appendix

Table 7-1: Different nodes and their Description:

node	Gene Description
abcA	ABC transporter, permease/ATP-binding protein
AgrA	accessory gene regulator protein A (autoinducer sensor protein response regulator protein)
AgrB	accessory gene regulator protein B (putative autoinducer processing protein)
AgrC	accessory gene regulator protein C (autoinducer sensor protein)
AgrD	accessory gene regulator protein D (AIP precursor)
arlR	DNA-binding response regulator ArlR
arlS	sensor histidine kinase ArlS
asp23	alkaline shock protein 23
atlA	bifunctional autolysin
aur	zinc metalloproteinase aureolysin
ccpa	catabolite control protein A
clfA	clumping factor A (fibrinogen and keratin binding surface anchored protein)
clfB	clumping factor B (fibrinogen and keratin binding surface anchored protein)
ClpP	locus for proteolytic subunit ClpP and the Clp ATPase ClpX
ClpX	locus for proteolytic subunit ClpP and the Clp ATPase ClpX
coa	staphylocoagulase precursor
cody	transcription pleiotropic repressor codY
fnbA	fibronectin binding protein A
fnbB	fibronectin binding protein B
geh	glycerol ester hydrolase
hla	alpha-haemolysin
hlb	beta-haemolysin
hld	delta-haemolysin
hlgA	gamma-haemolysin component A
hlgB	gamma-haemolysin component B
hlgC	gamma-haemolysin component C
icaA	intercellular adhesion protein A
icaB	intercellular adhesion protein B
icaC	intercellular adhesion protein C
icaD	intercellular adhesion protein D
icaR	ica operon transcriptional regulator
isaA	immunodominant antigen A
lrgA	holin-like protein LrgA
lrgB	holin-like protein LrgB
lytM	peptidoglycan hydrolase
lytN	cell wall hydrolase
lytR	two-component response regulator lytR
lytS	two-component sensor histidine kinase LytS
msa	hypothetical protein
msrR	peptide methionine sulfoxide reductase regulator MsrR
norA	multi drug resistance protein (norA)
rbf	ribosome-binding factor
Rot	repressor of toxins

6 – Appendix

RsbU	sigma factor B regulator protein
RsbV	anti-sigma B factor antagonist
RsbW	anti-sigmaB factor
SaeR	DNA-binding response regulator SaeR
SaeS	sensor histidine kinase SaeS
sak	staphylokinase precursor
SarA	staphylococcal accessory regulator A
SarR	staphylococcal accessory regulator R
SarS	staphylococcal accessory regulator S
SarT	staphylococcal accessory regulator T
SarU	staphylococcal accessory regulator U
SarV	staphylococcal accessory regulator V
SarX	staphylococcal accessory regulator X
SarZ	staphylococcal accessory regulator Z
sdrC	Ser-Asp rich fibrinogen-binding, bone sialoprotein-binding protein
SigB	RNA polymerase sigma-B factor
spa	immunoglobulin G binding protein A
SplA	serine protease SplA
SplB	serine protease SplB
SplC	serine protease SplC
SplD	serine protease SplD
SplE	serine protease SplE
SplF	serine protease SplF
ssaA	secretory antigen precursor
sspA	serine protease (V8 protease)
sspB	cysteine protease precursor
sspC	cysteine protease
tsst	toxic shock syndrome toxin-1

Listed here are all the nodes, used for the network, discussed in this thesis and their description.

Table 7-2: (A, B, C) Comparative microarray analysis:

A)

	agrA⁺ vs. agrA⁻								
node	A1: Cassat agrA OD 1	A1-T1 correlation	agrA up/down sim T1	A2: Cassat agrA OD 3	A2-T3 correlation	agrA up/down sim T3	A3: Dunman agrA RN27	A3-T3 correlation	agrA up/down sim T3
abcA	=		=	=		=	=		=
AgrA	x		x	x		x	x		x
AgrB	=		+	+		+	+		+
AgrC	=		=	=		=	=		=
AgrD	=		+	+		+	+		+
arlR	=		=	=		=	=		=
arlS	=		=	=		=	=		=
asp23	=		=	=		=	=		=
atlA	=		=	=		=	=		=
aur	=		=	=		+	+		+
ccpa	=		=	=		=	=		=
clfA	=		=	+		=	=		=
clfB	=		=	=		=	=		=
ClpP	=		=	=		=	=		=
ClpX	=		+	=		=	=		=
coa	=		=	=		=	=		=
cody	=		=	=		=	=		=
fnbA	=		+	+		+	=		+
fnbB	=		=	=		=	=		=
geh	=		+	+		+	+		+
hla	=		+	+		+	+		+
hlb	=		=	=		+	=		+
hld	+		+	+		+	+		+
hlgA	=		+	=		+	+		+
hlgB	=		+	=		+	+		+
hlgC	=		+	=		+	+		+
icaA	=		=	=		=	=		=
icaB	=		=	=		=	=		=

6 - Appendix

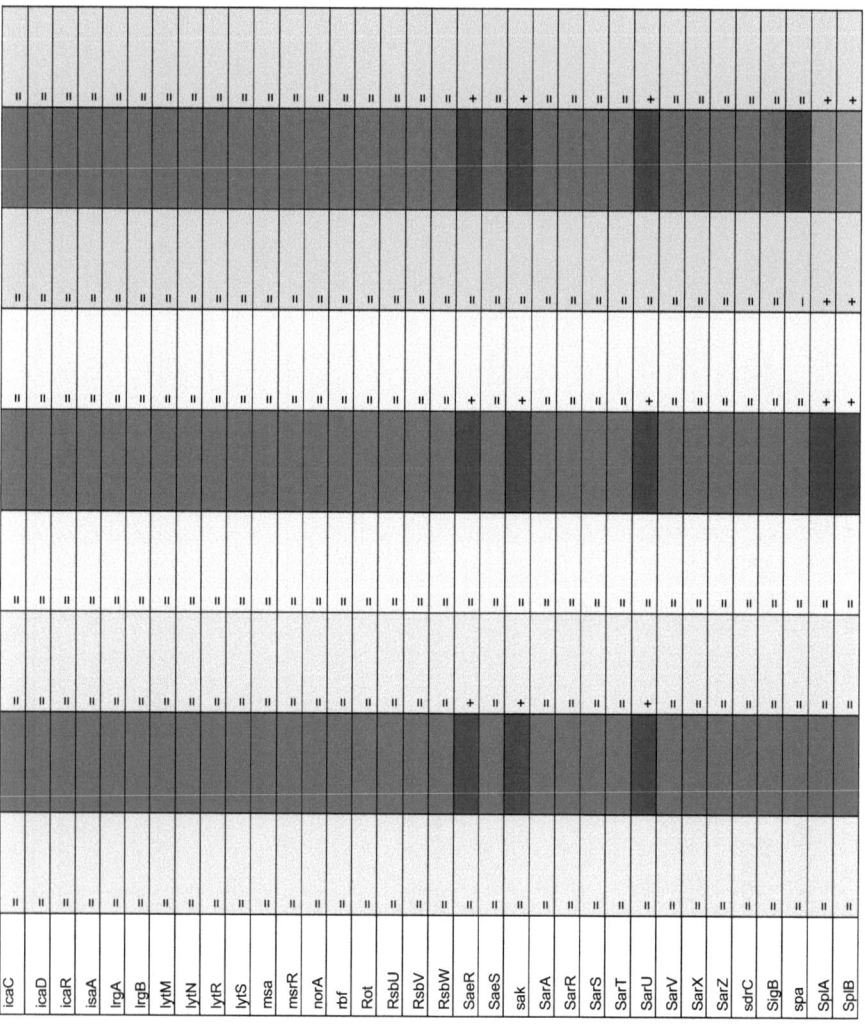

6 - Appendix

SplC	=	=		=		+		=
SplD	=	=		=		+	+	+
SplE	=	=		=		+	=	+
SplF	=	=		=		+	+	+
ssaA	=	=		=		=	=	=
sspA	=	=		=		+	=	+
sspB	=	=		=		=	+	=
sspC	=	=		=		=	=	=
tsst	=	=		=		=	=	=
Number of Nodes	70					70		70
concordant nodes	57					50		56
non concordant nodes	13					20		14
nodes with changes in the same direction	1					3		11
in vitro in silico consistency in %	81.43					71.43		80.00

93

6 - Appendix

B) node	sarA⁺ vs. sarA⁻ B1: Cassat sarA OD 1	B1-T1 correlation	sarA up/down sim T1	B2: Cassat sarA OD 3	B2-T3 correlation	sarA up/down sim T3	B3: Dunman agrA RN27	B3-T3 correlation	sarA up/down sim T3
abcA	=		=	=		=	=		=
AgrA	=		=	=		=	+		=
AgrB	=		+	=		=	+		=
AgrC	=		=	=		=	+		=
AgrD	=		+	=		=	+		=
arlR	=		=	=		=	=		=
arlS	=		=	=		=	=		=
asp23	=		=	=		=	–		=
atlA	–		=	–		=	–		=
aur	=		=	=		=	=		=
ccpa	=		=	=		=	=		=
clfA	=		=	=		=	=		=
clfB	=		=	=		=	=		=
ClpP	=		=	=		=	=		=
ClpX	=		=	=		=	=		=
coa	=		=	=		=	=		=
cody	=		=	=		=	=		=
fnbA	=		=	=		+	+		+
fnbB	=		+	=		+	=		+
geh	=		=	=		=	+		=
hla	–		=	=		+	+		+
hlb	=		=	=		=	=		=
hld	=		=	=		+	=		+
hlgA	=		+	=		+	+		+
hlgB	=		+	=		+	+		+
hlgC	=		+	=		+	+		+
icaA	=			=			=		
icaB	=			=			=		

6 - Appendix

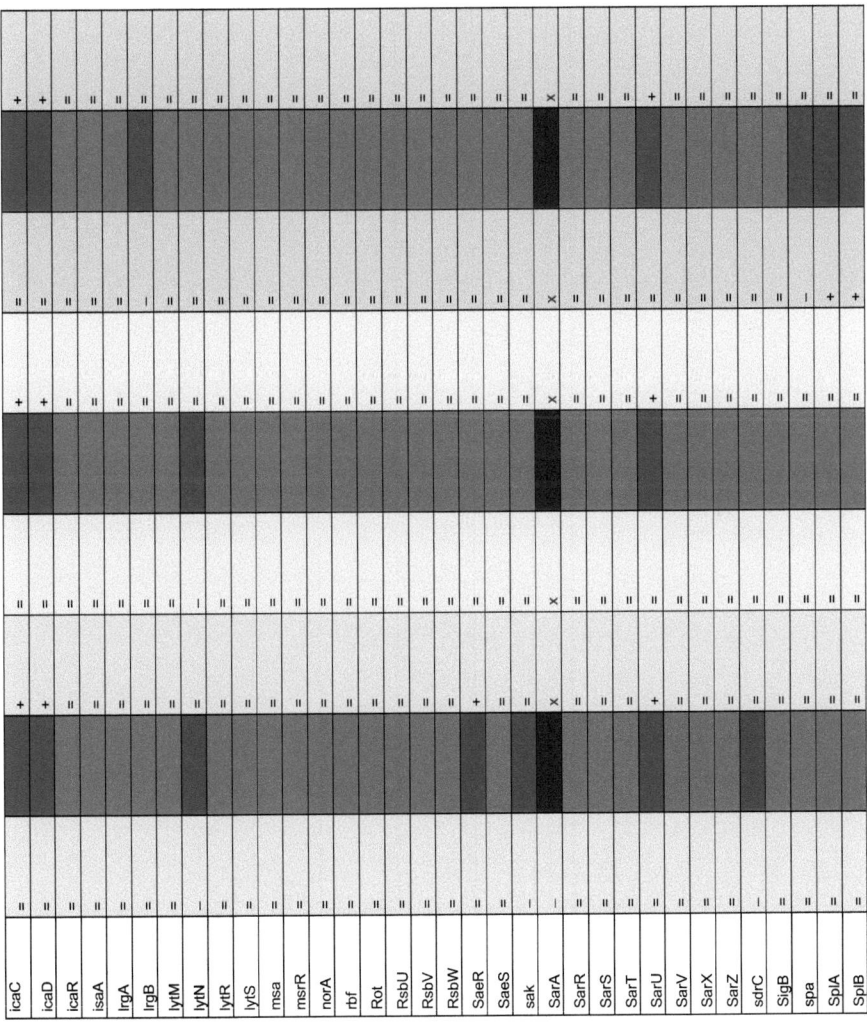

6 - Appendix

SplC	=	=	=	=	=	=
SplD	=	=	=	+	=	=
SplE	=	=	=	=	=	=
SplF	=	=	=	=	=	=
ssaA	=	=	=	=	=	=
sspA	−	−	=	+	=	+
sspB	−	−	=	=	=	=
sspC	−	−	=	=	−	=
tsst	=	+	=	+	=	+
Number of nodes		70		70		70
concordant nodes		50		53		48
non concordant nodes		20		17		22
nodes with changes in the same direction		0		0		4
in vitro in silico consistency in %		71.43		75.71		68.57

C) Biofilm vs. Planctonic

node	Biofilm Vs Planctonic (SS) Sim. 1	Biofilm Vs Planctonic (AIP) Sim. 2 (compared to Sim 1)	C1-Sim 1 correlation	C1 Sim 2 correlation	C1 Biofilm Vs Planctonic (maturing)	C2-Sim 1 correlation	C2-Sim 2 correlation	C2 Biofilm Vs Planctonic 24hr	C3-Sim 1 correlation	C3-Sim 2 correlation	C3 Biofilm Vs Planctonic OD 3.5
abcA	=	=			=			=			=
AgrA	=	=			=			=			=
AgrB	-	-			=			=			=
AgrC	-	=			=			=			=
AgrD	-	-			=			=			=
arlR	=	=			=			=			=
arlS	=	=			=			=			=
asp23	=	=			+			+			=
atlA	-	=			=			=			=
aur	=	=			=			=			=
ccpa	=	=			=			=			=
clfA	=	-			=			+			=
clfB	+	=			=			+			=
ClpP	=	=			=			=			=
ClpX	=	=			=			=			=
coa	=	=			=			=			=
cody	=	=			=			=			=
fnbA	=	=			=			=			=
fnbB	=	=			+			=			=
geh	-	=			+			=			=
hla	-	=			=			=			=
hlb	-	=			=			=			=
hld	-	-			=			=			=
hlgA	-	=			=			=			=
hlgB	-	=			=			=			=
hlgC	-	=			=			=			=
icaA	=	=			=			=			=

6 - Appendix

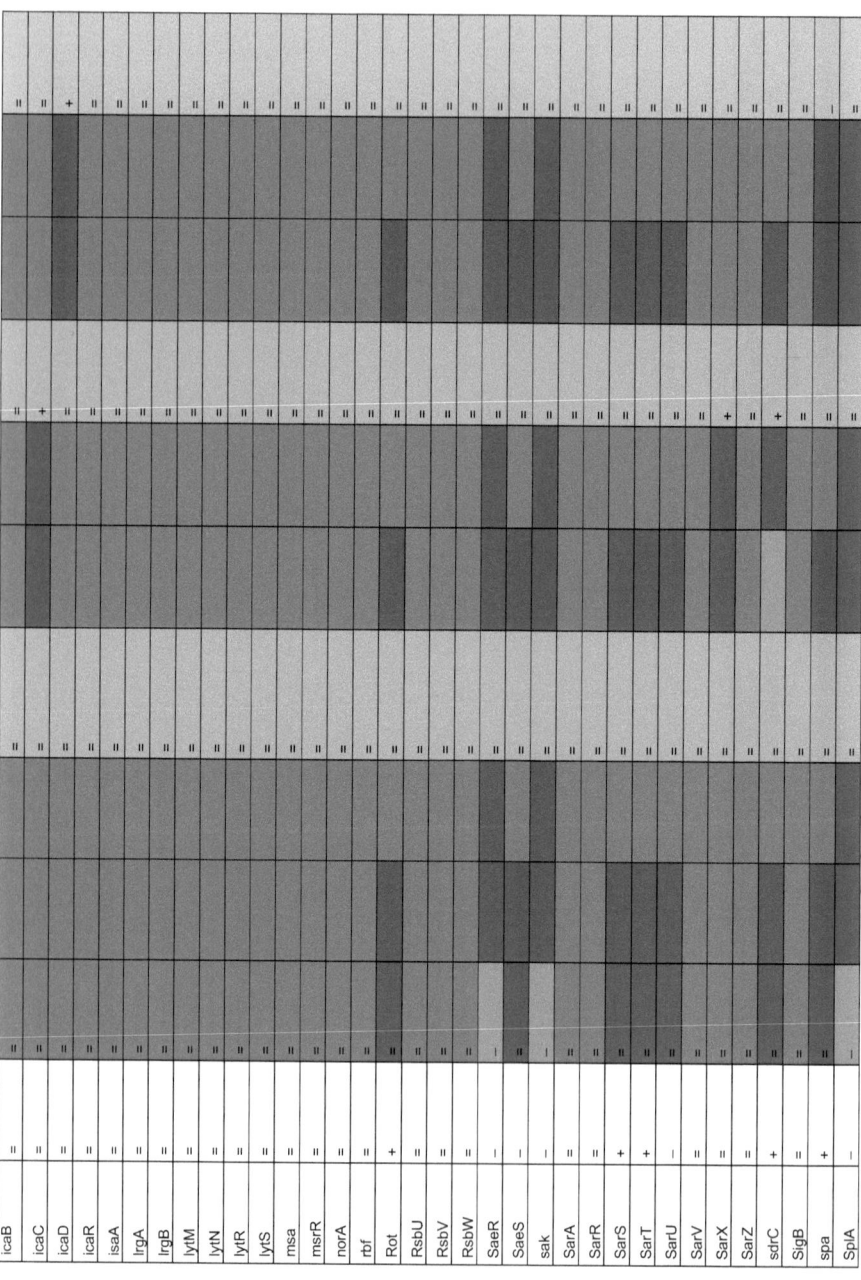

98

SplB	−	−	=	=	−
SplC	−	−	=	=	−
SplD	−	−	=	=	−
SplE	−	−	=	=	−
SplF	−	=	=	=	+
ssaA	=	=	=	=	=
sspA	=	=	=	=	=
sspB	=	=	=	=	=
sspC	=	=	=	=	=
tsst	=	=	=	=	=
Number of Nodes	71	71	71	71	71
concordant nodes	41	54	40	41	55
non concordant nodes	30	17	31	30	16
nodes with changes in the same direction	0	0	2	1	1
in vitro in silico consistency in %	57.75	76.06	56.34	57.75	77.46

	71	71
	41	55
	30	16
	1	1
	57.75	77.46

In (a) three *in vitro* $AgrA^+$ vs. $AgrA^-$ scenarios are compared to an *in silico* $AgrA^+$ vs. $AgrA^-$ scenario. Then in (b) three $SarA^+$ vs. $SarA^-$ scenarios are compared to an *in silico* $SarA^+$ vs. $SarA^-$ scenario. Furthermore in (c) three *in vitro* biofilm forming vs. not biofilm forming scenarios are compared to two *in silico* biofilm forming vs. not biofilm forming scenarios. A black shading in one of the correlation columns shows that this node was not included in the analysis, because this node is the one that was changed externally to get the different scenarios. The intermediate gray shading in one of the correlation columns means that this node showed no difference between *in vitro* and *in silico*. The dark gray in one of the correlation columns means that this node did not show the same reaction in the *in vitro* and the *in silico* situation. All other shadings are just for better visualisation of the compared groups. In this table a "+" means that this node is up-regulated by the three fold in the wild type strain or by the 2.5 fold in the biofilm forming situation. A "−" means that this node is up-regulated by the three fold in the mutant strain or by the 2.5 fold in the not biofilm forming situation. A detailed description of the scenarios can also be found in the Materials and Methods [chapter 2-2].

6 - Appendix

Table 7-3: Devices and materials used in this thesis:

Devices and Materials

Device/Material	Type	Company/location
Petri dishes		Greiner Bio-One; Frickenhausen; Germany
Test-tubes		Schott; Mainz; Germany
Photometer	Ultrospec 2100 pro	Amersham Bioscience/GE-Healthcare; Fairfield, Connecticut; USA
37°C shaker	HT-Infors	Infors; Bottmingen; Switzerland
One-way pipette		Corning Incorporated; Corning, New York; USA
Pipettor	pipetus	Hirschmann Laborgeräte; Herrenberg; Germany
Pipette		Eppendorf; Hamburg; Germany
Pipette tip		Greiner Bio-One; Frickenhausen; Germany
1.5ml cups		Eppendorf; Hamburg; Germany
1.5ml screw cups		Eppendorf; Hamburg; Germany
Plastic tubes 50ml; 20ml		Becton Dickinson; Franklin Lakes, New Jersey; USA
Centrifuge	Multifuge 3 SR	ThermoScientific; Waltham, Massachusetts; USA
Fastprep Shaker	Fastprep	MP Biomedicals; Irvine, California; USA
Vortexer	Julabo Paramix 3	Julabo Labortechnik; Seelbach; Germany
Heating Block	QBD 2	Grant Instruments; Cambridge; UK
Nylon membrane	Biodyne	Pall Corporation; Port Washington, New York; USA
Blotting papers	quickdraw	Sigma-Aldrich; St. Louis, Missouri; USA
Cross linker	GS Gene Linker	Bio-Rad; Hercules, California; USA
Rotator oven		MWG Biotech; Ebersberg; Germany
24-well plate		Greiner Bio-One; Frickenhausen; Germany
Blotter	Turbo Blotter	Schleicher and Schuell; Dassel; Germany
Weighing Scale	Kern 510	Kern und Sohn; Balingen-Frommern; Germany
Magnetic stirrer	Ikamag RCT	Ika; Staufen; Germany
Saranfoil	Saran	Dow; Midland, Michigan; USA
Film	Agfa Curix HT1000 G Plus Folienfilm	Agfa; Mortsel; Belgium

Dark hood Fotosystem		Biostep; Jahnsdorf; Germany
Blot Power Supply	Power Supply Model 1000/500	Bio-Rad; Hercules, California; USA
Sterile filters 0.22μm	Millex-HA, Syringe driven Filter	Merck, Millipore; Billerica; MA, USA
Cell Culture Flask	CellStar	Greiner Bio-One; Frickenhausen; Germany

Listed here are all materials, used fort the experiments in this thesis, including all the corresponding companies.

6 - Appendix

Table 7-4: Buffers and solutions used in this thesis:

Buffers and Solutions

Name	Composition		Company/location	Notes
CYPG	Casamino acid	10g	Becton Dickinson; Franklin Lakes, New Jersey; USA	
	Yeast	10g	Becton Dickinson; Franklin Lakes, New Jersey; USA	
	NaCl	5g	Merck; Darmstadt; Germany	
	H_2O	935ml		
	Glucose	25ml	AppliChem; Darmstadt; Germany	
	1.5M Glycerophosphat	40ml	AppliChem; Darmstadt; Germany	
Probe buffer	Formamid	750µl	Ambion/Life Technologies; Grand Island, NY; USA	
	10x MOPS	150µl	AppliChem; Darmstadt; Germany	
	Formaldehyde	262µl	AppliChem; Darmstadt; Germany	
	Ethidiumbromid (10mg/ml)	5µl	AppliChem; Darmstadt; Germany	
10x MOPS	MOPS (3-(N-morpholino) propanesulfonic acid)	40.5g	AppliChem; Darmstadt; Germany	
	NaAc	4.1g	Ambion; Grand Island, NY; USA	
	EDTA	20ml	AppliChem; Darmstadt; Germany	
	NaOH	4ml	Merck; Darmstadt; Germany	
	H20	1000ml		
transfer buffer	NaCl	175.5g	Merck; Darmstadt; Germany	
	Sarkosyl	0.62g	Sigma-Aldrich; St. Louis, Missouri; USA	
	Natriumhydroxid	0.32g	Merck; Darmstadt; Germany	
	H_2O	1000ml		

6 - Appendix

5x Phosphate buffer	Na$_2$HPO$_4$	79.25g	*AppliChem; Darmstadt; Germany*
	NaH$_2$PO$_4$	60.25g	*AppliChem; Darmstadt; Germany*
	H$_2$O	1000ml	
Hybridisation buffer High SDS			
	20xSSC	125ml	*AppliChem; Darmstadt; Germany*
	10xBlocking	100ml	Roche; Mannheim; Germany
	SDS (sodium dodecyl sulfate)	35g	*AppliChem; Darmstadt; Germany*
	10%-ige Laurylsarcosine	0.5ml	*AppliChem; Darmstadt; Germany*
	5x Phosphate buffer	25ml	*AppliChem; Darmstadt; Germany*
	H$_2$O	500ml	
10xPBS (Phosphate buffered saline)			
	NaCl	170g	Merck; Darmstadt; Germany
	Na$_2$HPO$_4$ x 12 H$_2$O	28.46g	Merck; Darmstadt; Germany
	KH$_2$PO$_4$	2.7g	Merck; Darmstadt; Germany
	H$_2$O	2000ml	
Bradford Solution			
	Coomassie Brilliant Blue G-250	60mg	Sigma-Aldrich; St. Louis, Missouri; USA
	3% Perchlorsäure	1000ml	Merck; Darmstadt; Germany
DNAse-Buffer			
	MgCl$_2$	12.2g	Merck; Darmstadt; Germany	60mM
	CaCl$_2$ (1M)	10ml	Merck; Darmstadt; Germany	10mM
	NaCl (0.14M)	590ml	Merck; Darmstadt; Germany	82.6mM
	Tris (1M)	400ml	LifeTechnologies; Carlsbad, CA, USA	400mM
Carbolgentianviolett			
	Phenol	0.5g	CarlRoth; Karlsruhe; Germany
	H$_2$O	20ml	

6 - Appendix

TAE	Ethanol	2ml	Merck; Darmstadt; Germany
	Cristalviolett	0.4g	Merck; Darmstadt; Germany
	Tris (1M)	40ml	LifeTechnologies; Carlsbad; CA, USA
	EDTA (0.5M)	2ml	Merck; Darmstadt; Germany
	Acetic acide	1ml	CarlRoth; Karlsruhe; Germany
	H_2O	957ml	
GelRed			Biotium, Inc; Hayward, CA; USA
BSA Standard				BCA Protein Assay Kit (Thermo Scientific)
			Thermo Scientific; Waltham; MA, USA
TSB (Tryptic Soy Broth)			Oxoid; Basingstoke; UK
EDTA 0.5M (Ethylendiamin-tetraacetat)			Merck; Darmstadt; Germany
Washing buffer			Roche; Mannheim; Germany	DIG wash and block buffer set (Roche)
Blocking solution			Roche; Mannheim; Germany	DIG wash and block buffer set (Roche)
Antibodies			Roche; Mannheim; Germany	Anti DIG AP (Roche)
CSPD			Roche; Mannheim; Germany	DIG wash and block buffer set (Roche)
Detection buffer			Roche; Mannheim; Germany	DIG wash and block buffer set (Roche)
Maleic acid			Roche; Mannheim; Germany	DIG wash and block buffer set (Roche)
Trizol			Invitrogen/LifeTechnologies; Carlsbad; CA, USA
Blue Juice			Invitrogen/LifeTechnologies; Carlsbad; CA, USA

Listed here are all the buffers and solutions, used fort the experiments in this thesis, their composition as well as the corresponding companies.

6 – Appendix

Table 7-5: Chemicals used in this thesis:

Chemicals	
Name	Company/location
Erythromycin	*BioChemica; Buchs; Switzerland*
Kanamycin	*AppliChem; Darmstadt; Germany*
Tetracycline	Serva Feinbiochemica; Heidelberg; Germany
Zirconiumsilicia-Beats	CarlRoth; Karlsruhe; Germany
Chloroform	*AppliChem; Darmstadt; Germany*
Isopropanol	*AppliChem; Darmstadt; Germany*
Ethanol	Merck; Darmstadt; Germany
1mM Natriumcitrat	Ambion/Life Technologies; Grand Island, NY; USA
Nuclease free water	Ambion/Life Technologies; Grand Island, NY; USA
Agarose	Biozym Scientific; Hessisch Oldendorf; Germany
Formaldehyde	*AppliChem; Darmstadt; Germany*
Glucose	*AppliChem; Darmstadt; Germany*
50% Methanol	Merck; Darmstadt; Germany
NaCl	Merck; Darmstadt; Germany
DNAse (Deoxyribonuclease I)	*AppliChem; Darmstadt; Germany*
Phenol (Aqua-Rothi-Phenol)	CarlRoth; Karlsruhe; Germany
Coomassie Brilliant Blue G	Sigma-Aldrich; St. Louis, Missouri; USA
H_2SO_4	Merck; Darmstadt; Germany
EDTA	Merck; Darmstadt; Germany
HCl	CarlRoth; Karlsruhe; Germany
H_3PO_4	CarlRoth; Karlsruhe; Germany

Listed here are all the chemicals, used fort the experiments in this thesis, including all the corresponding companies.

6 - Appendix

Table 7-6: Figures shown in this thesis:

Figure	Description	Source
Figure 1-1	Schematic, showing the *QS* in principle	own picture
Figure 1-2	*QS* around the luciferase operon in *V. fishery*	own picture, modified from Waters et al. 2005 [116]
Figure 1-3	Different Autoinducers	own picture, modified from Waters et al. 2005 [116]
Figure 1-4	*QS* around the *agr*-locus in *SA*	own picture
Figure 3-1	The two Steady states of the simulated network	own data and picture; figure already shown in Audretsch et al. 2013
Figure 3-3	Impact of *agr⁻* on the network	own data and picture; figure already shown in Audretsch et al. 2013
Figure 3-4	Reaction of different nodes, when knocking-out *agr* in silico	own data and picture; figure already shown in Audretsch et al. 2013
Figure 3-5	Impact of *saeRS⁻* on the network	own data and picture; figure already shown in Audretsch et al. 2013
Figure 3-6	Reaction of different nodes, when knocking-out *saeRS* in silico	own data and picture; figure already shown in Audretsch et al. 2013
Figure 3-7	Northern blot simulations.	own data and picture; figure already shown in Audretsch et al. 2013
Figure 3-8	Northern blot	own data and picture; figure already shown in Audretsch et al. 2013
Figure 3-9	Biofilm adherence assay	own picture
Figure 3-10	Venn-diagrams comparing genes, differentially expressed in *wt* vs. *sae⁻* and under planctonic vs. biofilm conditions	own data and picture
Figure 3-11	*DNAse* effect on the biofilm building ability of *wt* and *sae⁻* strains in silico	own data and picture
Figure 3-12	*DNAse* production in vitro	own data and picture
Figure 3-13	*DNAse* production in silico	own data and picture
Figure 3-14	Effect of different *DNAse*/NaCl solutions on biofilms of different *SA* strains	own data and picture
Figure 3-15	Effect of *DNAse*, DNAse-Buffer, H2O and NaCl on different *SA* biofilms	own data and picture

Figure 3-16	Amount of nucleic acids in different biofilms, qualitatively evaluated by blotting	own data and picture
Figure 4-1	Mathmatical simulation of the *agr*-locus	Jabbari et al. 2009 [51]
Figure 4-2	Schematic view of major apoptosis pathways in mammalian cells	Philippi et al. 2009 [83]
Figure 4-3	Growth of EPS^+ and EPS^- strains in biofilm culture	Bassler et al. 2011 [72]
Figure 4-4	Biofilm structures rendered from confocal micrograph stacks	Bassler et al. 2011 [72]
Figure 4-5	Colonizing bacterial biovolume after dispersal from biofilms	Bassler et al. 2011 [72]

Listed here are all the figures, shown in this thesis, their description as well as the corresponding source.

7 – References

7 - References

[1] R.P. Adhikari; R.P. Novick (2008). **Regulatory organization of the staphylococcal sae locus.** *Microbiology* **154**:949-959.

[2] A. Agarwal, K.P. Singh, A. Jain *(2010)*. **Medical significance and management of staphylococcal Biofilm.** *FEMS Immunol Med Microbiol* **58**:147-160.

[3] H. Akiyama; M. Ueda; H. Kanzaki; J. Tada; J. Arata *(1997)*. **Biofilm formation of *Staphylococcus aureus* strains isolated from impetigo and furuncle: role of fibrinogen and fibrin.** *Journal of Dermatological Science* **16**:2-10.

[4] A. Ballal; B. Ray; A.C. Manna *(2009)*. ***sarZ*, a *sarA* Family Gene, Is Transcriptionally Activated by MgrA and Is Involved in the Regulation of Genes Encoding Exoproteins in *Staphylococcus aureus*.** *Journal of Bacteriology* **191(5)**:1656-1665.

[5] A.H. Bartlett; K.G. Hulten *(2010)*. **Staphylococcus aureus Pathogenesis Secretion Systems, Adhesins, and Invasins.** *The Pediatric Infectious Disease Journal* **29(9)**:860-861.

[6] K.E. Beenken, P.M. Dunman, F. McAleese, D. Macapagal, E. Murphy, S.J. Projan, J.S. Blevins, M.S. Smeltzer, *(2004)*. **Global gene expression in *Staphylococcus aureus* biofilms.** *Journal of Bacteriology.* **186(14)**:4665-4684.

[7] M. Bischoff; J. M. Entenza; P. Giachino *(2001)*. **Influence of a Functional *sigB* Operon on the Global Regulators *sar* and *agr* in *Staphylococcus aureus*.** *Journal of Bacteriology 183(17)*:5171-5179.

[8] B.R. Boles, A.R. Horswill *(2008)*. **agr-Mediated Dispersal of Staphylococcus aureus Biofilms.** *PLoS Pathog* **4(4)**:e1000052.

[9] E. Bonabeau, M. Dorigo, G. Theraulaz *(1999)*. **Swarm intelligence: from natural to artificial systems.** *OUP USA* (No. 1).

[10] M.M. Bradford *(1976)*. *A rapid and sensitive method for the quantitation of microgram quantities of protein utilizing the principle of protein-dye binding.* Anal. Biochem. **72**:248-254.

[11] D.J. Bradshaw, P.D. Marsh, G.K. Watson, C. Allison *(1997)*. **Oral anaerobes cannot survive oxygen stress without interacting with facultative/aerobic species as a microbial community.** *Letters in Applied Microbiology.* **25**:385-387.

[12] R.A. Brady, J.G. Leid, A.K. Camper, J.W. Costerton, M.E. Shirtliff *(2006)*. **Identification of Staphylococcus aureus proteins recognized by the antibody-mediated immune response to a biofilm infection.** *Infect Immun.* **74(6)**:3415-3426.

7 − References

[13] S. Bronner; H. Monteil; G. Prévost *(2004)*. **Regulation of virulence determinants in Staphylococcus aureus: complexity and applications.** *FEMS Microbiology Reviews* **28**:183-200.

[14] N.C. Caiazza; G.A. O'Toole (2003). **Alpha-Toxin Is Required for Biofilm Formation by *Staphylococcus aureus*.** *Journal of Bacteriology* **185(10)**:3214-3217.

[15] M.C. Callegan, M.C. Booth, B.D. Jett, M.S. Gilmore *(1999)*. **Pathogenesis of gram-positive bacterial endophthalmitis.** *Infect Immun* **67**: 3348-3356.

[16] J. Carlsson *(1997)*. **Bacterial metabolism in dental biofilms.** *Adv. Dent. Res.* **11(1)**:75-80.

[17] J. Cassat; P.M. Dunman; E. Murphy; S.J. Projan; K.E. Beenken; K.J. Palm; S. Yang; K.C. Rice; K.W. Bayles; M.S. Smeltzer *(2006)*. **Transcriptional profiling of a Staphylococcus aureus clinical isolate and its isogenic agr and sarA mutants reveals global differences in comparison to the laboratory strain RN6390.** *Microbiology* **152**:3075-3090.

[18] A.L. Cheung; A.S. Bayer; G. Zhang; H. Gresham; Y. Xiong *(2004)*. **Regulation of virulence determinants in vitro and in vivo in Staphylococcus aureus.** *FEMS Immunology and Medical Microbiology* **40**:1-9.

[19] A.L. Cheung; K.A. Nishina; M.P.T. Pous; S. Tamber *(2008)*. **The SarA protein family of *Staphylococcus aureus*.** *Int. J. Biochem Cell Biol.* **40(3)**:355-361.

[20] P. Chopraa, A. Kammab *(2006)*. **Engineering Life through Synthetic Biology** *In Silico Biology.* **6**:401-410.

[21] G.D. Christensen; W.A. Simpson; J.J. Younger; L.M. Baddour; F.F. Barrett; D.M. Melton; E.H. Beachey *(1985)*. **Adherence of Coagulase-Negative Staphylococci to Plastic Tissue Culture Plates: a Quantitative Model for the Adherence of Staphylococci to Medical Devices.** *Journal of Clinical Microbiology* **22(6)**:996-1006.

[22] J. W. Costerton, K. J. Cheng, G. G. Geesey, T. I. Ladd, J. C. Nickel, M. Dasgupta, T. J. Marrie *(1987)*. **Bacterial biofilms in nature and disease.** *Annu. Rev. Microbiol.* **41**:435-464.

[23] D. Cue; M.G. Lei; T.T. Luong; L. Kuechenmeister; P.M. Dunman; S. O'Donnell; S; Rowe; J.P. O'Gara; C.Y. Lee *(2009)*. **Rbf Promotes Biofilm Formation by *Staphylococcus aureus* via Repression of *icaR*, a Negative Regulator of *icaADBC*.** *Journal of Bacteriology* **191(20)**:6363-6373.

[24] R. Dewanti, A.C.L. Wong *(1995)*. **Influence of culture conditions on biofilm formation by *Escherichia coli* 0157:H7.** *Int. J. Food Microbiol.* **26**:147-64.

[25] A. Di Cara; A. Garg; G. De Micheli; I. Xenarios; L. Mendoza *(2007)*. **Dynamic simulation of regulatory networks using SQUAD.** *BMC Bioinformatics* **8**:462.

7 - References

[26] R.M. Donlan; J.W. Costerton *(2002)*. **Biofilms: Survival Mechanisms of Clinically Relevant Microorganisms.** *Clinical Microbiology Reviews* **15(2):**167-193.

[27] M. DuBois, K.A. Gilles, J.K. Hamilton, P.A. Rebers, F. Smith (2002). **Colorimetric Method for Determination of Sugars and Related Substances.** *Anal. Chem.* **28(3):**350-356.

[28] S. Dubrac; I. Gomperts-Boneca; O. Poupel; T. Msadek *(2007)*. **New Insights into the *WalK/WalR* (*YycG/YycF*). Essential Signal Transduction Pathway Reveal a Major Role in Controlling Cell Wall Metabolism and Biofilm Formation in Staphylococcus aureus.** *Journal of Bacteriology* **189(22):**8257-8269.

[29] P. Dufour, S. Jarraud, F. Vandenesch, T. Greenland, RP. Novick *(2002)*. **High genetic variability of the agr locus in Staphylococcus species.** *J. Bacteriol.* **184:**1180-86.

[30] P.M. Dunman; E. Murphy; S. Haney; D. Palacios; G. Tucker-Kellogg; S. Wu; E. L. Brown; R. J. Zagursky; D. Shlaes; S. J. Projan *(2001)*. **Transcription Profiling-Based Identification of *Staphylococcus aureus* Genes Regulated by the *agr* and/or *sarA* Loci.** *Journal of Bacteriology* **183(24):**7341.

[31] W.M.Dunne, E.M Burd *(1992)*. **The effects of magnesium, calcium, EDTA and pH on the in vitro adhesion of Staphylococcus epidermidis to plastic.** *Microbiol. Immunol.* **36:**1019-1027.

[32] J,M. Entenza; P. Moreillon; M. M. Senn; J. Kormanec; P.M. Dunman; B. Berger-Bächi; S. Projan; M; Bischoff *(2005)*. **Role of σB in the Expression of *Staphylococcus aureus* Cell Wall Adhesins *ClfA* and *FnbA* and Contribution to Infectivity in a Rat Model of Experimental Endocarditis.** *Infection and Immunity* **73(2):**990-998.

[33] B. Fournier; R. Aras; D.C. Hooper *(2000)*. **Expression of the Multidrug Resistance Transporter *NorA* from *Staphylococcus-aureus* Is Modified by a Two-Component Regulatory System.** Journal of Bacteriology **182(3):**664-671.

[34] B. Fournier; A. Klier; G. Rapoport (2001). **The two-component system *ArlS-ArlR* is a regulator of virulence gene expression in Staphylococcus aureus.** *Molecular Microbiology* **41(1):**247-261.

[35] A. Funahashi, Y. Matsuoka, A. Jouraku, M. Morohashi, N. Kikuchi, H. Kitano *(2008)*. **Cell Designer 3.5: A versatile modelling tool for biochemical networks.** *Proceedings of the IEEE* **96:**1254-1265.

[36] T. Geiger; C. Goerke; M. Mainiero; D. Kraus; C. Wolz *(2008)*. **The Virulence Regulator Sae of *Staphylococcus aureus*: Promoter Activities and Response to Phagocytosis-Related Signals.** *Journal of Bacteriology* **190(10):**3419-3428.

[37] S.R. Gill; D.E. Fouts; G.L. Archer; E.F. Mongodin; R.T. DeBoy; J. Ravel; I.T. Paulsen; J.F. Kolonay; L. Brinkac; M. Beanan; R.J. Dodson; S.C. Daugherty; R. Madupu; S.V. Angiuoli; A.S. Durkin; D.H. Haft; J. Vamathevan; H. Khouri; T. Utterback; C. Lee; G. Dimitrov; L. Jiang; H. Qin; J. Weidman; K. Tran; K.

7 – References

Kang; I.R. Hance; K.E. Nelson; C.M. Fraser *(2005)*. **Insights on Evolution of Virulence and Resistance from the Complete Genome Analysis of an Early Methicillin-Resistant *Staphylococcus aureus* Strain and a Biofilm-Producing Methicillin-Resistant *Staphylococcus epidermidis* Strain.** *J. Bacteriol.* **187(7)**:2426-2438.

[38] A.T. Giraudo; A.L. Cheung; R. Nagel *(1997)*. **The sae locus of Staphylococcus aureus controls exoprotein synthesis at the transcriptional level.** *Arch. Microbiol.* **168**:53-58.

[39] A.T. Giraudo; A. Calzolari; A.A. Cataldi; C. Bogni; R. Nagel (1999). **The *sae* locus of *Staphylococcus aureus* encodes a two-component regulatory system.** *FEMS Microbiology Letters* **177**:15-22.

[40] C. Goerke; U. Fluckiger; A. Steinhuber; V. Bisanzio; M. Ulrich; M. Bischoff; J.M. Patti; C. Wolz *(2005)*. **Role of *Staphylococcus aureus* Global Regulators *sae* and σB in Virulence Gene Expression during Device-Related Infection.** *Infect. Immun.* **73(6)**:3415-3421.

[41] C.A. Gordon, N.A. Hodges, C. Marriott *(1988)*. **Antibiotic interaction and diffusion through alginate and exopolysaccharide of cystic fibrosisderived Pseudomonas aeruginosa.** *J Antimicrob Chemother* **22**:667-74.

[42] F. Götz *(2002)*. **Staphylococcus and Biofilms.** Molecular Microbiology **43(6)**:1367-1378.

[43] E. Gustafsson, P. Nilsson, S. Karlsson, S. Arvidson *(2004)*. **Characterizing the Dynamics of the Quorum-Sensing System in Staphylococcus aureus.** *J Mol Microbiol Biotechnol* **8**:232-242.

[44] E. Gustafsson; J. Oscarsson *(2008)*. **Maximal transcription of aur (aureolysin) and *sspA* (serine protease) in Staphylococcus aureus requires staphylococcal accessory regulator R (*sarR*) activity.** *FEMS Microbiol Lett.* **284**:158-164.

[45] L. Hall-Stoodley, J.W. Costerton, P. Stoodley *(2004)*. **Bacterial Biofilms from the natural Environment to infectious diseases.** *Nature review, microbiology.* **2**:95-108.

[46] L. Hall-Stoodley, F.Z. Hu, A. Gieseke, L. Nistico, D. Nguyen, J. Hayes, M. Forbes, D.P. Greenberg, B. Dice, A. Burrows, P.A. Wackym, P. Stoodley, J.C. Post, G.D. Ehrlich, J.E. Kerschner *(2006)*. **Direct detection of bacterial biofilms on the middle-ear mucosa of children with chronic otitis media.** *JAMA.* **296**:202-211.

[47] M. Hausner, S. Wuertz *(1999)*. **High rates of conjugation in bacterial biofilms as determined by quantitative in-situ analysis.** *Appl. Environ. Microbiol.* **65**:3710-3713.

[48] K. Hiramatsu; H. Hanaki; T. Ino; K. Yabuta; T. Oguri; F. C. Tenover *(1997)*.

7 - References

Methicillin-resistant Staphylococcus aureus clinical strain with reduced vancomycin susceptibility. *Journal of Antimicrobial Chemotherapy* **40**:135-146.

[49] M.J. Huseby; A.C. Kruse; J. Digre; P.L. Kohler; J.A. Vocke; E.E. Mann; K.W. Bayles; G.A. Bohach; P.M. Schlievert; D.H. Ohlendorf; C.A. Earhart *(2010)*. **Beta toxin catalyzes formation of nucleoprotein matrix in staphylococcal Biofilms.** *Proc. Natl. Acad. Sci. U. S. A.* **107(32)**: 14407-14412.

[50] S.S. Ingavale; W. Van Wamel; A.L. Cheung *(2003)*. **Characterization of RAT, an autolysis regulator in *Staphylococcus aureus*.** *Molecular Microbiology* **48(6)**:1451-1466

[51] S. Jabbari; J.R. King; A.J. Koerber; P. Williams *(2009)*. **Mathematical modelling of the *agr* operon in *Staphylococcus aureus*.** *J. Math. Biol.*

[52] T. Jin; M. Bokarewa; T. Foster; J. Mitchell; J. Higgins; A. Tarkowski *(2004)*. ***Staphylococcus aureus* Resists Human Defensins by Production of Staphylokinase, a Novel Bacterial Evasion Mechanism.** *The Journal of Immunology* **172**:1169-1176.

[53] M. Johnson; A. Cockayne; J.A. Morrissey *(2008)*. **Iron-Regulated Biofilm Formation in *Staphylococcus aureus* Newman Requires *ica* and the Secreted Protein Emp.** *Infection and Immunity* **76(4)**:1756-1765.

[54] V. Joly, B. Pangon, J.M. Vallois, L. Abel, N. Brion, A. Bure, N.P. Chau, A. Contrepois, C. Carbon. *(1987)*. **Value of antibiotic levels in serum and cardiac vegetations for predicting antibacterial effect of ceftriaxone in experimental Escherichia coli endocarditis.** *Antimicrob. Agents Chemother.* **31**:1632-1639.

[55] A. Karlsson; P. Saravia-Otten: K. Tegmark; E. Morfeldt; S. Arvidson *(2001)*. **Decreased Amounts of Cell Wall-Associated Protein A and Fibronectin-Binding Proteins in *Staphylococcus aureus sarA* Mutants due to Up-Regulation of Extracellular Proteases.** *Infection and Immunity* **69(8)**:4742-4748.

[56] S. Kjelleberg, S. Molin *(2002)*. **Is there a role for quorum sensing signals in bacterial biofilms?** *Curr Opin Microbiol.* **5(3)**:254-8.

[57] S. Klamt, J. Saez-Rodriguez, J.A. Lindquist, L. Simeoni, E.D. Gilles *(2006)*. **A methodology for the structural and functional analysis of signalling and regulatory networks.** *BMC Bioinformatics* **7**:56.

[58] C. Koch, N. Hoiby *(1993)*. **Pathogenesis of cystic fibrosis.** *Lancet.* **341**:1065-1069.

[59] J. Krumsiek, S. Pölsterl, D.M. Wittmann, F.J. Theis *(2010)*. **Odefy-from discrete to continuous models.** *BMC Bioinformatics* **11**:233.

[60] H. Kuroda; M. Kuroda; L. Cui; K. Hiramatsu *(2007)*. **Subinhibitory concentrations**

of beta-lactam induce haemolytic activity in *Staphylococcus aureus* through the *SaeRS* two-component system. *FEMS Microbiol Lett.* **268(1)**:98-105.

[61] A.J. Koerber, J.R. King, P. Williams *(2005)*. **Deterministic and stochastic modelling of endosome escape by Staphylococcus aureus: "quorum sensing" by a single bacterium.** *J Math Biol* **50**:440-488.

[62] C.S. Laspidou, B.E. Rittmann, *(2002)*. **A unified theory for extracellular polymeric substances, soluble microbial products, and active and inert biomass.** *Water Research.* **(36)**:2711-2720.

[63] J.R. Lawrence, G.D.W. Swerhone1, G.G. Leppard, T. Araki, X. Zhang, M.M. West, A.P. Hitchcock *(2003)*. **Scanning Transmission X-Ray, Laser Scanning, and Transmission Electron Microscopy Mapping of the Exopolymeric Matrix of Microbial Biofilms.** *Appl. Environ. Microbiol.* **69(9)**:5543-5554.

[64] G.Y. Liu, A.Essex, J.T. Buchanan, V. Datta, H.M. Hoffman, J.F. Bastian, J. Fierer, V. Nizet *(2005)*. **Staphylococcus aureus golden pigment impairs neutrophil killing and promotes virulence through its antioxidant activity.** *The Journal of Experimental Medicine.* **202(2)**:209-215.

[65] F.D. Lowy *(1998)*. **Staphylococcus aureus infections.** *The New England Journal of Medicine* **339(8)**:520-532.

[66] M. Mainiero; C. Goerke; T. Geiger; C. Gonser; S. Herbert; C. Wolz *(2010)*. **Differential Target Gene Activation by the *Staphylococcus aureus* Two-Component System *saeRS*.** *Journal of Bacteriology* **192(3)**:613-623.

[67] C.D. Majerczyk; M.R. Sadykov; T.T. Luong; C. Lee; G.A. Somerville; A.L. Sonenshein (2008). **Staphylococcus aureus CodY Negatively Regulates Virulence Gene Expression.** *Journal of Bacteriology* **190(7)**:2257-2265.

[68] A.C. Manna; A.L. Cheung *(2003)*. **sarU, a sarA Homolog, Is Repressed by SarT and Regulates Virulence Genes in Staphylococcus aureus.** *Infection and Immunity* **71(1)**:343-353.

[69] L. Mendoza, I. Xenarios *(2006)*. **A method for the generation of standardized qualitative dynamical systems of regulatory networks.** *Theor Biol Med Model.* **3**:13.

[70] M.B. Miller; B.L. Bassler *(2001)*. **Quorum sensing in Bacteria.** *Annu. Rev. Microbiol.* **55**:165-99.

[71] C.P. Montgomery; S. Boyle-Vavra; R.S. Daum *(2010)*. **Importance of the Global Regulators Agr and SaeRS in the Pathogenesis of CA-MRSA USA300.** *Infection. PLoS ONE.* **5(12)**:e15177.

[72] C.D. Nadella, B.L. Bassler *(2011)*. **A fitness trade-off between local competition and dispersal in Vibrio cholerae biofilms.** *PNAS.* **108(34)**:14181-14185.

7 - References

[73] M. Nagata; C. Kaito; K. Sekimizu *(2008)*. **Phosphodiesterase Activity of *CvfA* Is Required for Virulence in *Staphylococcus aureus*.** *The Journal of Biological Chemistry* **283(4)**:2176-2184.

[74] K.H. Nealson, J.W. Hastings *(1979)*. **Bacterial bioluminescence: its control and ecological significance.** *Microbiol. Rev.* **43**:496-518.

[75] N.N. Nickerson; L. Prasad; L. Jacob; L.T. Delbaere; M.J. McGavin *(2007)*. **Activation of the SspA Serine Protease Zymogen of *Staphylococcus aureus* Proceeds through Unique Variations of a Trypsinogen-like Mechanism and Is Dependent on Both Autocatalytic and Metalloprotease-specific Processing.** *The Journal of Biological Chemistry* **282(47)**:34129-34138.

[76] R.P. Novick, H.F. Ross, S.J. Projan, J. Kornblum, B. Kreiswirth, S Moghazeh. *(1993)*. **Synthesis of staphylococcal virulence factors is controlled by a regulatory RNA molecule.** *EMBO J.* **12**:3967-75.

[77] R.P. Novick *(2003)*. **Autoinduction and signal transduction in the regulation of staphylococcal virulence.** *Molecular Microbiology* **48(6)**:1429-1449.

[78] R.P. Novick; D. Jiang *(2003)*. **The staphylococcal saeRS system coordinates environmental signals with agr quorum sensing.** *Microbiology.* **149**:2709-2717.

[79] M. Palma; A. Haggar; J.I. Flock *(1999)*. **Adherence of *Staphylococcus aureus* is enhanced by an endogenous secreted protein with broad binding activity.** *Journal of Bacteriology* **181(9)**:2840-2845.

[80] M. Palma; A.L. Cheung *(2001)*. **σB Activity in *Staphylococcus aureus* Is Controlled by *RsbU* and an Additional Factor(s) during Bacterial Growth.** *Infection and Immunity* **69(12)**:7858-7865.

[81] M. Palma; A. Bayer; L.I. Kupferwasser; T. Joska; M.R. Yeaman; A.L. Cheung *(2006)*. **Salicylic Acid Activates Sigma Factor B by *rsbU*-Dependent and -Independent Mechanisms.** *Journal of Bacteriology* **188(16)**:5896-5903.

[82] P.A. Pattee *(1981)*. **Distribution of Tn551 Insertion Sites Responsible for Auxotrophy on the Staphylococcus aureus Chromosome.** *Journal of Bacteriology* **145(1)**:479-488.

[83] N. Philippi; D. Walter; R. Schlatter; K. Ferreira; M. Ederer; O. Sawodny; J. Timmer; C. Borner; T. Dandekar *(2009)*. **Modelling system states in liver cells: Survival, apoptosis and their modifications in response to viral infection.** *BMC Systems Biology* **3**:97.

[84] L.A. Pratt, R. Kolter *(1998)*. **Genetic analysis of *Escherichia coli* biofilm formation: defining the roles of flagella, motility, chemotaxis and type I pili.** *Mol. Microbiol.* **30(2)**:285-94.

[85] N. Qureshi, B.A. Annous, T.C. Ezeji, P. Karcher, I.S. Maddox *(2005)*. **Biofilm reactors for industrial bioconversion processes: employing potential of**

7 – References

enhanced reaction rates. *Microbial Cell Factories.* **4(24)**.

[86] A. Resch, R. Rosenstein, C. Nerz, F. Gotz, *(2005).* **Differential gene expression profiling of Staphylococcus aureus cultivated under biofilm and planktonic conditions.** *Appl Environ Microbiol.* **71(5)**:2663-2676.

[87] K.C. Rice; T. Patton; S.J. Yang; A. Dumoulin; M. Bischoff; K.W. Bayles *(2004).* **Transcription of the *Staphylococcus aureus* cid and *lrg* Murein Hydrolase Regulators Is Affected by Sigma Factor B.** *Journal of Bacteriology* **186(10)**:3029-3037.

[88] K.C. Rice; E.E. Mann; J.L. Endres; E.C. Weiss; J.E. Cassat; M.S. Smeltzer; K.W. Bayles *(2007).* **The *cidA* murein hydrolase regulator contributes to DNA release and Biofilm development in Staphylococcus aureus.** *Proc. Natl. Acad. Sci. U. S. A.* **104(19)**:8113-8118.

[89] K. Rogasch; V. Rühmling; J. Pané-Farré; D. Höper; C. Weinberg; S. Fuchs; M. Schmudde; B.M. Bröker; C. Wolz; M. Hecker; S. Engelmann *(2006).* **Influence of the Two-Component System *SaeRS* on Global Gene Expression in Two Different *Staphylococcus aureus* Strains.** *Journal of Bacteriology.* **188(22)**:7742-7758.

[90] R. Rosenstein, C. Nerz, L. Biswas, A. Resch, G. Raddatz, S.C. Schuster; F. Götz *(2009).* **Genome Analysis of the Meat Starter Culture Bacterium *Staphylococcus carnosus* TM300.** *Appl. Environ. Microbiol.* **75(3)**:811.

[91] J. Rossi; M. Bischoff; A. Wada; B. Berger-Bächi *(2003).* **MsrR, a Putative Cell Envelope-Associated Element Involved in *Staphylococcus aureus sarA* Attenuation.** *Antimicrobial Agents and Chemotherapy* **47(8)**:2558-2564.

[92] H.L. Saenz, V. Augsburger, C. Vuong, R.W. Jack, F. Gotz, M. Otto *(2000).* **Inducible expression and cellular location of AgrB, a protein involved in the maturation of the staphylococcal quorum-sensing pheromone.** *Arch. Microbiol.* **174**:452-55.

[93] P. Melke; P. Sahlin; A. Levchenko; H. Jönsson *(2010).* **A Cell-Based Model for Quorum Sensing in Heterogeneous Bacterial Colonies.** *PLoS Comput. Biol.* **6(6)**:e1000819.

[94] B. Saïd-Salim; P.M. Dunman; F.M. McAleese; D. Macapagal; E. Murphy; P.J. McNamara; S. Arvidson; T.J. Foster; S.J. Projan; B.N. Kreiswirth *(2003).* **Global Regulation of *Staphylococcus aureus* Genes by Rot.** *Journal of Bacteriology* **185(2)**:610-619.

[95] K. Sambanthamoorthy; M.S. Smeltzer; M.O. Elasri *(2006).* **Identification and characterization of *msa* (SA1233), a gene involved in expression of *SarA* and several virulence factors in Staphylococcus aureus.** *Microbiology* **152**:2559-2572.

[96] R. Schlatter, N. Philippi, G. Wangorsch, R. Pick, O. Sawodny, C. Borner, J. Timmer,

M. Ederer, T. Dandekar *(2011)* **Integration of Boolean models exemplified on hepatocyte signal Transduction.** *Briefings in Bioinformatics.* **13(3)**:365-376.

[97] K.A. Schmidt; A.C. Manna; S. Gill; A.L. Cheung *(2001).* **SarT, a Repressor of α-Haemolysin in *Staphylococcus aureus*.** *Infection and Immunity* **69(8)**:4749-4758.

[98] K.A. Schmidt; A.C. Manna; A.L. Cheung *(2003).* **SarT Influences *sarS* Expression in *Staphylococcus aureus*.** *Infection and Immunity* **71(9)**:5139-5148.

[99] J. Schneider, A. Yepes, J. C. Garcia-Betancur, I. Westedt, B. Mielich, D. **Streptomycin-Induced Expression in Bacillus subtilis of YtnP, a Lactonase-Homologous Protein That Inhibits Development and Streptomycin Production in Streptomyces griseus** López, *Appl Environ Microbiol.* 2012, **78(2)**, 599-603.

[100] C.U. Schwermer, G. Lavik, R.M.M. Abed, B. Dunsmore, T.G. Ferdelman, P. Stoodley, A. Gieseke, D. de Beer *(2008).* **Impact of Nitrate on the Structure and Function of Bacterial Biofilm Communities in Pipelines Used for Injection of Seawater into Oil Fields.** *Applied and environmental microbiology.* **74(9)**:2841-2851.

[101] P.C. Seed, L. Passador, B.H. Iglewski *(1995).* **Activation of the *Pseudomonas aeruginosa lasI* gene by LasR and the *Pseudomonas* autoinducer PAI: an autoinduction regulatory hierarchy.** *J. Bacteriol.* **177**:654-59.

[102] K. Seidl; C. Goerke; C. Wolz; D. Mack; B. Berger Bächi; M. Bischoff *(2008).* ***Staphylococcus aureus* CcpA Affects Biofilm Formation.** *Infection and Immunity* **76(5)**:2044-2050.

[103] U. Szewzyk, R. Szewzyk *(2003).* **Biofilme - die etwas andere Lebensweise.** *BIOspektrum* **3/03(9)**:253-255.

[104] K.J. Tack, L.D. Sabath *(1985).* **Increased minimum inhibitory concentrations with anaerobiasis for tobramycin, gentamicin, and amikacin, compared to latamoxef, piperacillin, chloramphenicol, and clindamycin.** *Chemotherapy* **31**:204-10.

[105] A. Toledo-Arana; N. Merino; M. Vergara-Irigaray; M. Débarbouillé; J.R. Penadés; I. Lasa *(2005).* ***Staphylococcus aureus* Develops an Alternative, *ica*-Independent Biofilm in the Absence of the *arlRS* Two-Component System.** *Journal of Bacteriology* **187(15)**:5318.

[106] .O.Toole, H.B. Kaplan, R. Kolter *(2000).* **Biofilm formation as microbial development.** *Annu. Rev. Microbiol.* **54**:49-79.

[107] M.P. Trotonda; A.C. Manna; A.L. Cheung; I. Lasa; J.R. Penadés *(2005).* **SarA Positively Controls Bap-Dependent Biofilm Formation in *Staphylococcus aureus*.** *Journal of Bacteriology* **187(16)**:5790-5798.

[108] Q.C. Truong-Bolduc; J. Strahilevitz; D.C. Hooper *(2006).* **NorC, a New Efflux Pump**

7 – References

Regulated by *MgrA* of *Staphylococcus aureus*. *Antimicrobial Agents and Chemotherapy* **50(3)**:1104-1107.

[109] Q.C. Truong-Bolduc; D.C. Hooper *(2007)*. **The Transcriptional Regulators NorG and *MgrA* Modulate Resistance to both Quinolones and β-Lactams in *Staphylococcus aureus*.** *Journal of Bacteriology 189(8)*:2996-3005.

[110] E. Tuomanen, R. Cozens, W. Tosch, O. Zak, A. Tomasz (1986). **The rate of killing of Escherichia coli by β-lactam antibiotics is strictly proportional to the rate of bacterial growth.** *J Gen Microbiol* **132**:1297-304.

[111] K.L. Visick, J. Foster, J. Doino, M. McFall-Ngai, E.G. Ruby *(2000)*. ***Vibrio fischeri* lux genes play an important role in colonization and development of the host light organ.** *J. Bacteriol.* **182**:4578-86.

[112] C. Vuong; F. Götz; M. Otto *(2000)*. **Construction and Characterization of an *agr* Deletion Mutant of *Staphylococcus epidermidis*.** *Infect. Immun.* **68(3)**:1048-1053.

[113] C. Vuong; H.L. Saenz; F. Götz; M. Otto *(2000)*. **Impact of the *agr* Quorum-Sensing System on Adherence to Polystyrene in *Staphylococcus aureus*.** *The Journal of Infectious Diseases* **182**:1688-93.

[114] J.P. Ward; J.R. King; A.J. Koerber; P. Williams; J.F. Croft; R.E. Sockett *(2001)*. **Mathematical modelling of quorum sensing in bacteria.** *IMA Journal of Mathematics Applied in Medicine and Biology* **18**:263-292.

[115] C. Walsh *(2000)*. **Molecular mechanisms that confer antibacterial drug resistance.** *Nature* **406**:775-781.

[116] C.M.Waters, B.L. Bassler *(2005)*. **Quorum Sensing: Cell-to-Cell Communication in Bacteria.** *Annu. Rev. Cell Dev. Biol.* **21**:319-46.

[117] P.I. Watnick, K.J. Fullner, R. Kolter *(1999)*. **A role for the mannose-sensitive hemagglutinin in biofilm formation by *Vibrio cholerae* El Tor.** *J. Bacteriol.* **181(11)**:3606-9.

[118] C. Wolz; P. Pöhlmann-Dietze; A. Steinhuber; Y.T. Chien; A. Manna; W. van Wamel; A.L. Cheung *(2000)*. ***Agr*-independent regulation of fibronectin-binding protein(s) by the regulatory locus *sar* in *Staphylococcus aureus*.** *Molecular Microbiology* **36(1)**:230-243.

[119] S.J. Yang; K.C. Rice; R.J. Brown; T.G. Patton; L.E. Liou; Y.H Park; K.W. Bayles *(2005)*. **A LysR-Type Regulator, CidR, Is Required for Induction of the *Staphylococcus aureus cidABC* Operon.** *Journal of Bacteriology* **187(17)**:5893-5900.

[120] J.M. Yarwood; P.M. Schlievert *(2003)*. **Quorum sensing in Staphylococcus infections.** *J. Clin. Invest.* **112**:1620-1625.

[121] J.M. Yarwood; D.J. Bartels; E.M. Volper; E.P. Greenberg *(2004)*. **Quorum Sensing**

7 - References

in *Staphylococcus aureus* Biofilms. *Journal of Bacteriology* **186(6)**:1838-1850.

[122] R.N. Zadoks, J.R. Middleton, S. McDougall, J. Katholm, Y.H. Schukken *(2011)*. **Molecular Epidemiology of Mastitis Pathogens of Dairy Cattle and Comparative Relevance to Humans.** *J. Mammary Gland Biol. Neoplasia.* **16**:357-372.

[123] W. Ziebuhr, V. Krimmer, S. Rachid, I. Lobner, F. Gotz, J. Hacker *(1999).* **A novel mechanism of phase variation of virulence in *Staphyococcus epidermidis*: evidence for control of the polysaccharide intercellular adhesion synthesis by alternating insertion and excision of the insertion sequence element IS*256*.** *Mol. Microbiol.* **32(2)**:345-56.

i want morebooks!

Buy your books fast and straightforward online - at one of world's fastest growing online book stores! Environmentally sound due to Print-on-Demand technologies.

Buy your books online at
www.get-morebooks.com

Kaufen Sie Ihre Bücher schnell und unkompliziert online – auf einer der am schnellsten wachsenden Buchhandelsplattformen weltweit! Dank Print-On-Demand umwelt- und ressourcenschonend produziert.

Bücher schneller online kaufen
www.morebooks.de

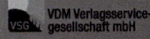 VDM Verlagsservicegesellschaft mbH
Heinrich-Böcking-Str. 6-8   Telefon: +49 681 3720 174   info@vdm-vsg.de
D - 66121 Saarbrücken   Telefax: +49 681 3720 1749   www.vdm-vsg.de

Printed by Books on Demand GmbH, Norderstedt / Germany